Multiple Criteria Decision Making

Series Editor

Constantin Zopounidis, School of Production Engineering and Management, Technical University of Crete, Chania, Greece

This book series focuses on the publication of monographs and edited volumes of wide interest for researchers and practitioners interested in the theory of multicriteria analysis and its applications in management and engineering. The book series publishes novel works related to the foundations and the methodological aspects of multicriteria analysis, its applications in different areas in management and engineering, as well as its connections with other quantitative and analytic disciplines. In recent years, multicriteria analysis has been widely used for decision making purposes by institutions and enterprises. Research is also very active in the field, with numerous publications in a wide range of publication outlets and different domains such as operations management, environmental and energy planning, finance and economics, marketing, engineering, and healthcare.

This series has been accepted by Scopus.

Konstantina Ragazou • Constantin Zopounidis •
Alexandros Garefalakis • Nikolaos Sariannidis

Triple Bottom Line and Multiple Criteria Decision Making Analysis

Fuzzy Selection Processes in the Financial Sector

Springer

Konstantina Ragazou
Department of Management
Science and Technology
University of Western Macedonia
Kozani, Greece

Department of Business Administration
Neapolis University Pafos
Pafos, Cyprus

Constantin Zopounidis
School of Production Engineering and
Management
Financial Engineering Laboratory, Technical
University of Crete
Chania, Greece

Alexandros Garefalakis
Department of Business Administration
and Tourism
Hellenic Mediterranean University
Heraklion, Greece

Nikolaos Sariannidis
Department of Accounting and Finance
University of Western Macedonia
Kozani, Greece

ISSN 2366-0023 ISSN 2366-0031 (electronic)
Multiple Criteria Decision Making
ISBN 978-3-031-78044-8 ISBN 978-3-031-78045-5 (eBook)
https://doi.org/10.1007/978-3-031-78045-5

This Springer imprint is published by the registered company Springer Nature Switzerland AG
The registered company address is: Gewerbestrasse 11, 6330 Cham, Switzerland

Preface

The significance of environmental sustainability in the business world is on the rise, necessitating that companies transfer their focus from evaluating and comparing their environmental impact to identifying the most effective methods for reducing it. Most financial institutions' decisions are significantly influenced by financial metrics. However, it is imperative to evaluate indicators from both the environmental and social domains as sustainability becomes a commercial objective. Nevertheless, conducting such an assessment in a pragmatic and methodical manner is a difficult task. We offer an implementation of the multi-criteria decision analysis (MCDA) technique that addresses the identified challenges within the financial institution framework. As a result, we employ financial calculation methodologies and company surveys to evaluate environmental, economic, and social performance indicators, respectively. Weight entropy, fuzzy analysis, and TOPSIS methodologies have been employed to develop a comprehensive framework for organizations with a variety of stakeholders. The company's leadership can achieve successful implementation by prioritizing the selection of solutions from a variety of stakeholders. The organization effectively incorporated a variety of indicators that were consistent with its social and environmental responsibilities by employing MCDA methodologies. This approach resulted in a comprehensive examination of the alternatives that considered factors beyond economics. We have discovered that the inclusion of triple bottom line indicators in our MCDA application provides a more comprehensive understanding of stakeholder preferences. Consequently, our book has established a framework for structured discussions regarding the organization's objectives and priorities, as well as a comprehensive comprehension of the available alternatives.

Kozani, Greece — Konstantina Ragazou
Chania, Greece — Constantin Zopounidis
Heraklion, Greece — Alexandros Garefalakis
Kozani, Greece — Nikolaos Sariannidis

Preface

Contents

Chapter 1
Triple Bottom Line and Decision-Making: An Investigation into the Concept of Confluence

1.1 How Does the Triple Bottom Line Work?

Over the last several years, sustainability has become a significant objective for governments, organizations, and enterprises. Evaluating the extent to which a company is actively pursuing sustainable growth or implementing sustainability initiatives may be challenging. In the mid-1990s, John Elkington developed a novel framework to assess the performance of corporations in the United States with regard to sustainability. The triple bottom line (TBL) accounting system expanded the scope of conventional metrics, including profitability, return on investment (ROI), and shareholder value, by incorporating environmental and social concerns. Triple bottom line reporting is an effective strategy for achieving sustainability objectives by prioritizing comprehensive investment outcomes, which include financial performance, social welfare, and environmental consequences [1, 2].

The use of TBL accounting has become more popular in the for-profit, nonprofit, and government sectors. Multiple philanthropic organizations and companies have used the TBL sustainability framework to assess their operational effectiveness. Furthermore, governments at the federal, state, and local levels have used a similar strategy [3]. This chapter evaluates the concept of triple bottom line (TBL); highlights its potential benefits for specialists in economic development, governments, and businesses; and showcases recent examples of TBL application. Moreover, it delves into the role of TBL in the decision-making process [3].

1.1.1 An Explanation of the Triple Bottom Line (TBL)

The triple bottom line (TBL) is an accounting term that encompasses three key performance indicators: financial, environmental, and social. This approach diverges

K. Ragazou et al., *Triple Bottom Line and Multiple Criteria Decision Making Analysis*, Multiple Criteria Decision Making,
https://doi.org/10.1007/978-3-031-78045-5_1

from traditional reporting frameworks by including ecological and social aspects, which may complicate the identification of appropriate quantification methods. The triple bottom line (TBL) is frequently characterized by three pillars: people, environment, and economic advantages. These will be classified as the three Ps. Before Elkington introduced the concept of sustainability as the "triple bottom line," environmentalists were struggling to determine the methods and frameworks for evaluating sustainability [4, 5]. Over the last three decades, there has been a substantial rise in the number of academic fields dedicated to the study of sustainability. Andrew Savitz's thorough explanation of the triple bottom line (TBL) is supported by sustainability professionals, regardless of their affiliation with academia. The triple bottom line (TBL) idea quantifies the impact of an organization's actions on the global community by considering its financial performance, shareholder interests, and social, human, and environmental resources. The difficulty is precisely in defining TBL. The critical factor in measurement is precision [6].

1.1.1.1 The Background of the Triple Bottom Line

The concept of sustainability began to acquire momentum in the 1980s. The Brundtland Commission of the United Nations devised the concept of sustainability in 1987. They defined it as the ability to satisfy current needs without compromising the ability of future generations to meet their own needs. The environmental, social, and economic dimensions were considered when formulating this concept. The Commission specifically prioritized natural resources when evaluating sustainability. It is alleged that humanity must implement significant modifications to its current consumption practices, which currently regard natural resources as infinite [6, 7].

Additionally, it is essential to challenge the current trend of causing damage to our own ecosystem. It is imperative to also consider the social structures that establish the foundation within which we can thrive and develop. Brundtland's concept of sustainability posits that our social institutions and environmental systems need care and restoration due to their limited flexibility [8, 9]. In relation to the Brundtland Commission, our economic systems serve as the unifying factor that unites all of these components. The long-term survival of civilization is contingent upon the existence of systems that are both resilient and healthy, which is the essence of true sustainability.

The components provide the fundamental framework for the triple bottom line, which is the assessment and quantification of the financial, environmental, and social returns on investment. A company must achieve authentic sustainability, which is the ability to endure and generate profits for all its stakeholders, by achieving success in three critical areas of capital. Regarding the sustainability of businesses, a pertinent query that may arise considering the facts presented is: How does this affect businesses, and why should they be concerned? A business will cease to exist if humanity fails to persist. Nevertheless, the potential extinction of humanity is not a pressing or relevant issue for corporate executives, who are primarily focused on the upcoming financial quarter. It is highly unlikely that any management would allocate business time to deliberate on this issue, and it is reasonable for them to refrain from doing so [10, 11]. However, it is crucial for organizations to acknowledge that the issue is not inconsequential and should not be disregarded.

Although the threat to all life on Earth is undeniable, economic activities significantly exacerbate this issue. The corporate sector possesses a unique advantage in addressing this problem. A reorganization of the issue that is pertinent to the current aims and objectives of the company and that is comprehensible to a business expert is required [10]. Without a doubt, the primary objective of organizations in the current era is to achieve financial prosperity. The primary objective of organizations is to generate profit for their shareholders. However, this was not consistently accurate. Initially, corporations were established with the primary objective of achieving specific social objectives, such as the construction of essential infrastructure, such as roads, bridges, dams, or other critical structures [12, 13].

Additionally, the duration of these enterprises was frequently constrained. The term of a company's activities extended indefinitely to an infinite extent, and its primary objective became the pursuit of financial gain. Currently, the prioritization of financial profit is of the utmost importance, particularly within publicly listed corporations. To optimize immediate financial gains for current stakeholders, companies are required to operate within restricted time frames, typically concentrated on quarterly earnings [12]. This phenomenon is a result of the unwavering zeal to "optimize stockholder profits." Regrettably, in the present day, most shareholders are short-term speculators, as opposed to long-term investors, who prioritize the company's survival. They prioritize short-term gains over the company's long-term development. Many managers of public companies would acknowledge that these pressures may induce them to make decisions that are advantageous to shareholders in the immediate future but that have adverse long-term repercussions for the firm and all its stakeholders, including long-term investors [14]. To achieve sustainability and deliver enduring value to shareholders and other stakeholders, a company must divest its attention from its inordinate emphasis on quarterly profits and its short-term financial success. The process may commence with the revision of the management objectives, as directed by the board of directors, to redirect the emphasis from solely increasing "shareholder profits" to maximizing "stakeholder value." This new objective is more comprehensive and inclusive, thereby benefiting a broader range of individuals and groups. Investors, creditors, suppliers, consumers, the community, and the environment comprise the stakeholders. Many of these stakeholders frequently have deadlines that are both undefined and extended. Managers may optimize stakeholder value by extending the time frames and considering a broader range of perspectives during the planning and execution process. Consequently, the triple bottom line (TBL) enables the optimization of value for all stakeholders, including shareholders [14, 15].

1.2 The Three Bottom Line's Capital Components

The concept of the triple bottom line pertains to the assessment and quantification of the economic, ecological, and societal dimensions of a capital undertaking. This statement is based on the premise that there are three fundamental types of capital that are crucial to a company's success and should therefore be compensated with a "return on investment" [16, 17]. There are three distinct types of capital: economic capital, natural capital, and social capital.

1.2.1 Economic Capital

Businesses primarily acknowledge and accept only cash or economic capital, and our existing accounting systems are designed to reflect this acknowledgment. The significance of cash capital and the methods used to quantify it are widely recognized and will not be examined in this context [18]. It is widely acknowledged that for a firm to be sustainable, it must generate profits on its financial statement. The other types of capital, and its connection to sustainability from the business viewpoint, are not as comprehensively grasped and will be examined in that context.

1.2.2 Natural Capital

Natural capital encompasses the entirety of Earth's resources, including raw materials, flora, and fauna. Currently, we are depleting these resources at an escalating pace. Presently, we regard these resources as if they are inexhaustible, despite the evident fact that they are not [19]. Anyone with even a basic understanding of business understands that continuously spending cash capital without sufficient replenishment would inevitably result in the company's downfall. It is important to acknowledge that the planet's natural resources, known as natural capital, are also limited. We must not continue to deplete these resources without replenishing them. Businesses that are unable to get the necessary raw materials for production are unable to maintain their operations, like those who exhaust their financial resources [20].

An approach to tackle this problem is changing our current mindset of "cradle to grave" in relation to the manufacturing and selling of physical products. Corporations recognize that reducing waste directly leads to cost savings. It is important for them to acknowledge that waste is a valuable resource that should be recycled and reused rather than thrown away [21]. This strategy is commonly referred to as "cradle to cradle" to symbolize a self-sustaining cycle. Every organization that has embraced the cradle-to-cradle idea in its business practices has experienced substantial improvements in its financial performance.

1.2.3 Social Capital

Social capital, the third form of capital, is a subjective resource that is likely the most crucial contribution of capital. There are two sorts of social capital. The first component is the human capital contributed by individuals [22, 23] (workers, contractors, suppliers, and advisers) directly to the firm. The second factor is the allocation of resources by the social systems that aid the business [24–26].

1.2.3.1 Human Capital

Human capital refers to the valuable knowledge, skills, and abilities possessed by individuals, which they take with them when they leave work at the end of the day. Even after the initial financial investment has fulfilled its purpose, it is the intellectual and skill-based resources of individuals that continue to enter a company and support its operations. It refers to the individual commitment and dedication of individuals who devote their time and effort to ensure the success of a firm [26]. Human capital refers to the intellectual, creative, experiential, skilled, talented, passionate, and educated resources contributed by employees, their families, suppliers, partners, consultants, and customers. Employees are an essential asset in any organization. However, personnel are viewed as a cost, a replaceable asset (like inventory) that can be reduced in large numbers with the belief that this will result in improved financial performance. Massive labor cutbacks, when seen from the perspective of capital disposal, are as illogical as discarding dollar bills (a company's monetary capital) on a street corner [25, 26]. Several studies have unequivocally shown that companies that lay off a substantial proportion of their employees for the purpose of immediate cost reduction do not perform any better and often perform worse in the long run, compared to similar companies in the same industry that do not engage in mass layoffs. However, boards of directors, senior management, and their consultants continue to utilize this strategy in response to immediate market and financial demands. The pursuit of immediate financial benefits continues to be the primary motivation behind poor decision-making, leading to long-term consequences that impact the overall economic performance [27, 28].

Even in cases where these organizations do not need to contemplate reducing their personnel, there is a consistent tendency to not view employees as a form of capital, particularly one that should yield a return. Indeed, employees do receive pay, which includes salaries and bonuses. However, these forms of remuneration are typically proportional to the value they contribute and do not constitute a genuine return on investment [29]. Over the years, several stock schemes, such as option plans, have been created to tackle this problem. However, because of tax implications and considerable differences in how these benefits are distributed, most employees do not believe that they are included in these systems in a fair and equal manner. For most employees, the opportunity to have a fair and proportionate ownership stake in the company they work for is a difficult goal to achieve, leading to increasing dissatisfaction. The resentment is being fueled by highly imbalanced remuneration for senior executives and various other variables affecting the treatment of employees. As a result, turnover rates are higher. Employee discontent with their firm directly contributes to turnover, which in turn negatively impacts the financial bottom line. On the other hand, acknowledging, effectively preserving, and incentivizing this valuable workforce will result in improved financial performance.

1.2.3.2 Social System Capital

Prefecture capital investment in the social system is mostly derived from educational, governmental, and infrastructural institutions. These institutions play a crucial role in providing the necessary basis for the development and growth of corporations. In a community with a dysfunctional justice system or a lack of contract enforcement, limited transportation options such as mules instead of efficient highways or the Internet, an uneducated and unskilled workforce, an unreliable communication system, an underdeveloped financial system that cannot generate investment capital, and restrictive state regulations hindering business startups, it is challenging to envision the difficulties of establishing a successful business.

The present corporate landscape neglects to recognize the role of human capital, as well as the impact of the social system on capital. Social institutions depend on taxes, subsidies, and other contributions to ensure their financial sustainability, and they often seek aid as a last option. Nevertheless, corporations would be incapable of maintaining and growing without the financial investment that is permitted by their capital contribution. A more effective strategy is necessary to guarantee that the sources of social capital in a system provide a positive return on their investment. By adopting this method, companies may benefit from a workforce that is highly qualified and competent, customers who have more discretionary cash, and enhanced infrastructure, all of which contribute to their success and prosperity. These several aspects will enhance the financial success of each firm.

1.3 Measurement of the Triple Bottom Line

There is no universally recognized unit of measurement for the 3Ps. Currency serves as the unit of measurement for profits. Numerous indicators and metrics are frequently implemented to evaluate social capital. What is the current state of environmental and ecological health? It is a significant challenge to establish a unified standard for measurement. Advocates promote the commercialization of all elements of the triple bottom line (TBL), which encompasses environmental harm and social welfare. Although there are advantages to using dollars as a unified unit of measurement, there are those who are opposed to designating a monetary value to endangered species or wetlands solely on philosophical grounds. There are individuals who are skeptical of the methodology employed to determine the appropriate valuation for endangered species or depleted wetlands. Utilizing an index may serve as an alternative approach to computing the TBL. By resolving the issue of incompatible units, this method enables comparisons between a variety of entities, including enterprises, cities, development initiatives, or other benchmarks. This assumes the existence of a universally recognized accounting methodology.

A comparative measure of a county's performance across several components in relation to the entire country is the Innovation Index, which was devised by the Indiana Business Research Center. However, a certain degree of subjectivity persists, despite the utilization of an index. What is the methodology employed to assign weights to the index's constituents? Would each "P" be allocated the same weight? What are the various components that are contained within each "P"? Is the weighting distributed equally among each of them? Is the category of humanity more significant than the category of the planet? Who is accountable for the decision-making process? An alternative approach would involve refraining from employing currency units or indicators to assess sustainability. Each sustainability metric would be autonomous if the TBL's users possessed the courage and resilience to do so. By monitoring changes in the establishment, devastation, or preservation of wetlands over time, the term "acres of wetlands" may be employed to assess progress. The substantial proliferation of indicators that may be pertinent for sustainability assessment is a fundamental drawback of this methodology. Users may experience metric fatigue because of the triple bottom line (TBL) concept [30, 31]. After addressing the challenges associated with calculating the triple bottom line (TBL), our focus is now on selecting the most suitable metrics for use in a TBL calculation. Subsequently, we will investigate the implementation of the TBL concept by corporations and other organizations.

1.3.1 What Measures Go into the Index?

At present, there is no method that is universally agreed upon or explicitly defined for evaluating the total benefit of learning (TBL). Additionally, there is a lack of agreement regarding the universally acknowledged criteria for quantifying the components of the triple bottom line (TBL) [32, 33]. This feature is advantageous because it enables users to customize the framework to accommodate the requirements of a variety of organizations (e.g., nongovernmental organizations (NGOs) or enterprises), distinct projects or policies, or diverse geographical regions. The reduction of solid refuse disposed of in landfills is a common metric that both corporations and local government organizations may employ to assess environmental sustainability. However, a municipal mass transport system may evaluate its effectiveness by examining the number of kilometers traveled by passengers, whereas a for-profit bus operator may evaluate its performance by calculating the earnings per share. The TBL can accommodate these modifications. Additionally, the triple bottom line (TBL) can be tailored to suit specific cases or projects, enabling a more focused investigation of the geographic boundaries of a small town or a comprehensive analysis of the consequences across large geographical regions [32–34]. The effects of a singular undertaking in a defined geographic region, such as a community involved in the development of a park, would be assessed by a project-specific TBL. The triple bottom line (TBL) framework can be implemented at the state level to evaluate energy policy and infrastructure initiatives at the national level. The level, project type, and geographic scope of the business will have a substantial impact on the selection of metrics to include. However, the ultimate decision regarding the actions will be contingent upon the feasibility of gathering the requisite data and the perspectives of professionals and those with vested interests. Although there is a significant amount of research on the appropriate standards for evaluating sustainability at the local or national level, the computations for the triple bottom line (TBL) will ultimately depend on the availability of data. Scientific discussions have been conducted to rigorously scrutinize and verify the sustainability indicators that have been referenced [35].

Furthermore, there is no widely recognized approach for measuring the total bottom line (TBL). There is a lack of consensus over the exact criteria used to assess the indicators in the three areas of the triple bottom line (TBL). This feature is advantageous because it allows users to customize the framework to meet the needs of different entities, such as businesses, nonprofit organizations, and various projects or policies, like investments in infrastructure or educational programs. It also allows for customization based on different geographical boundaries, such as a city, region, or country. Both a local government body and a corporation may evaluate environmental sustainability by using similar criteria, such as minimizing the amount of solid waste sent to landfills [36, 37]. However, a local mass transport system may assess its efficiency by analyzing the overall distance covered by passengers, whereas a for-profit bus business can assess its performance by analyzing its income per share. The TBL can effectively handle these inconsistencies. In addition, the triple bottom line (TBL) may be customized to suit individual circumstances or projects, enabling a thorough examination of the impacts over large geographical areas or a more targeted assessment within a limited geographic scope, such as a small town. A case-specific TBL would evaluate the impacts of a particular project in a specific area, such as the establishment of a park by a community.

Furthermore, the TBL framework may be used to evaluate energy policy at a national scale or infrastructure efforts at the state level. The selection of metrics to include will mostly depend on the entity's level, the nature of the project, and the geographic extent. The selection of measurements will eventually be determined by those who have specialized expertise in the topic, a personal interest in the issue, and the ability to obtain the necessary data. The calculations for the triple bottom line (TBL) will depend on the availability of data, regardless of the extensive literature discussing the suitable criteria for assessing sustainability at the local or national levels.

Below are a few classic sustainability measures that have been thoroughly examined and approved through academic discourse (Table 1.1):

Table 1.1 Examples of indicators considered in the TBL's measurement

Economic	Environmental	Social
Individual earnings	Nitrogen oxide concentration	Unemployment rate
Expenses related to underemployment	Priority pollutant selection	Female labor force participation rate
Rate of turnover in businesses	Excessive nutrient levels	Median household income
Size of businesses	Electricity usage	Relative poverty
Increase in employment opportunities	Fossil fuel usage	Percentage of population with a postsecondary degree or certificate
Distribution of jobs across different industries	Management of solid waste	Average commute time
Proportion of businesses in each industry	Management of hazardous waste	Violent crimes per capita
Contribution of each industry's revenue to the overall state economy	Alteration in land use/land cover	Health-adjusted life expectancy

1. *Economic indicators*: the transfer of money and financial outcomes should be influenced by economic indicators. The research may incorporate variables such as income, expenditures, taxes, business environment, employment, and company diversity [38, 39].
2. *Environmental indicators*: to provide a precise representation of the factors that may affect the long-term sustainability of natural resources, environmental indicators should incorporate measurable data. Numerous variables, such as energy consumption, natural resources, solid and hazardous waste, land use/land cover, and air and water quality, can be integrated by the system. Ideally, companies would be able to evaluate the potential impact of a project or policy on the region by obtaining long-term trends for each environmental indicator [38, 40].
3. *Social indicators*: education, health and well-being, educational equity and accessibility to social services, quality of life, and social capital are among the diverse social aspects of a community or region that are comprised of social indicators [39].

A multitude of these indicators are gathered at the state and national levels, and they are also available for examination at the local or community level. A community is presented with an abundance of alternatives when developing a triple bottom line (TBL). Nevertheless, the appropriate measures may vary depending on the project's geographical scope and characteristics. The TBL contends that the most effective approach to evaluating success indicators for local or community-based initiatives is to incorporate local characteristics and circumstances.

Various strategies may be implemented to ensure that stakeholders are actively engaged and provide valuable input during the development of the TBL framework. These methods entail the creation of a decision matrix that incorporates public preferences into project planning and decision-making, the use of a "narrative format" to enable stakeholders to rank and prioritize components of a sustainability framework based on community priorities, and the promotion of stakeholder engagement and comprehensive project evaluation. For instance, a community may regard the number of firms owned by women that have been established over a 5-year period as a significant indicator of the effectiveness of an entrepreneurial development program. In the end, the organization will be accountable for the creation of a definitive set of metrics that are pertinent to the current endeavor.

1.3.2 Alternative Methods of Triple Bottom Line Measuring

The TBL framework is utilized by governments, businesses, and organizations to advance social, environmental, and economic sustainability. However, the methods by which they assess the results in these three domains differ. Proponents who have developed and implemented sustainability evaluation frameworks, such as the TBL, have encountered considerable difficulty, specifically in identifying suitable data for the components of the index and in devising a comprehensive and practical index. Twenty-five variables comprise the Genuine Progress Indicator (GPI), which encompasses social, economic, and environmental concerns. By converting the variables to monetary units, they are consolidated into a solitary metric that is expressed in dollars. For example, Minnesota has developed an internal progress indicator comprising 42 variables, which evaluate the state's advancement toward the goals of fostering a strong economy.

Integrated assessment and sustainability measures have been the subject of a considerable body of literature, with their origins in the disciplines that quantify environmental impacts. The assessments of alterations in social welfare are not constrained by an inflexible economic theory. Scholars specializing in environmental policy argue that a comprehensive understanding of the ramifications of a policy, regulation, or economic development endeavor necessitates the integration of economic, social, and environmental factors. This integration facilitates a thorough evaluation of policy alternatives and compromises.

1.4 Who Can Implement the TBL Approach?

Government, nonprofit organizations, and businesses can all apply the TBL approach.

1.4.1 Government

Government institutions at the state, regional, and municipal levels are increasingly using the triple bottom line (TBL) and similar sustainability assessment frameworks to evaluate performance and make decisions. Analyses have been performed in several locations, including Maryland, Minnesota, Vermont, Utah, the San Francisco Bay Area, and Northeast Ohio, utilizing the TBL or a similar sustainability paradigm. Policymakers use these sustainability evaluation frameworks to ascertain the optimal course of action for enhancing societal sustainability. In addition, policymakers strive to understand the cause-and-effect link between activities, such as projects or programs, and their impact on society's movement toward or away from sustainability. In addition, the State of Maryland utilizes a hybrid GPI-TBL framework to assess efforts, such as investments in renewable energy, by comparing them to the baseline of "inaction" or other policy options.

The European Union uses integrated assessment to determine the possible favorable and unfavorable outcomes of proposed policy initiatives on the worldwide level. This allows the development of informed political judgments on the planning and identification of the necessary compromises to reconcile opposing goals. EU laws have faced criticism and have undergone many rounds of reform. The European Union's dedication to comprehensive assessment is emphasized by the act of revising legislation, which also showcases the openness of the process.

1.4.2 Nonprofit Organizations

A multitude of charitable organizations have embraced the TBL approach and have formed alliances with private enterprises to tackle wide-ranging sustainability concerns that impact shared stakeholders. Businesses acknowledge the benefits of partnering with nonprofit groups, especially those that aim to promote economic growth, social welfare, and environmental conservation. The Ford Foundation has provided funding for studies that have utilized several iterations of the TBL framework to assess the impacts of initiatives aimed at enhancing economic prosperity in numerous rural areas throughout the United States. Another illustration is Resilience and Sustainability Facility (RSF) Social Finance, a nonprofit company that specifically concentrates on how its investments enhance all three dimensions of the TBL. RSF offers a unique perspective on the TBL concept, but it is evident that the TBL may be customized to suit almost any organization. Its methodology encompasses the following:

- Conduct an examination of innovative economic models that foster sustainable practices in the food and agriculture sector while concurrently raising public awareness of the significance of organic and biodynamic farming.
- Ecological stewardship (environmental): direct financial resources toward organizations and programs that are committed to preserving, revitalizing, and protecting the planet's ecosystems, with a specific emphasis on holistic, interconnected, and culturally suitable approaches.
- Education funds: direct financial resources toward educational and artistic programs that are all-encompassing and have a beneficial effect on mental and emotional well-being.

1.4.3 *Businesses*

The TBL, along with its fundamental principle of sustainability, has gained significant traction in the corporate sphere because of mounting anecdotal proof of enhanced long-term profitability. For instance, by minimizing the amount of waste generated by packaging, it is possible to concurrently decrease expenses. General Electric, Unilever, Proctor and Gamble, 3M, and Cascade Engineering are some of the companies that have been exemplary in implementing these strategies. While some companies may not have a TBL based on an index, it is possible to observe how they assess sustainability by employing the TBL framework.

One example of a private corporation, Cascade Engineering, which is not required to submit extensive financial documents like public companies, has determined the following criteria for its TBL scorecard:

- Economic
  - Tax burden
- Social
  - Mean duration of employee training
  - Transitioning from welfare to long-term employment
  - Philanthropic donations
- Environmental and safety concerns
  - Safety incident rate refers to the frequency of safety incidents that occur during a certain period of time.
  - Lost/restricted workday rate measures the number of workdays lost or limited due to injuries or illnesses.
  - Sales dollars per kilowatt hour is a metric that calculates the amount of sales generated per unit of energy produced.
  - Greenhouse gas emissions quantify the amount of gases, such as carbon dioxide, released into the atmosphere as a result of human activities.

1.5 Triple Bottom Line as a Decision-Making Tool

The TBL decision-making tool (or "the Tool") was created to assist staff in adhering to the TBL policy while strategizing new initiatives and modifying existing projects. Utilize the TBL decision-making tool to determine the effects of your effort on a set of success indicators and gain insight into areas where you may minimize trade-offs, negative consequences, and substantial detrimental effects. Both a thorough version and a condensed version of the TBL decision-making tool have been created. Both versions undergo annual reviews to ensure they integrate employee feedback, absorb changes to the organization, and adapt to industry best practices. The TBL decision-making tool is structured according to principles, indications, and success metrics.

Principles encompass advantages related to the environment, society, economy, and governance. Indicators are employed to delineate certain areas of concentration for each component of the TBL. Success measures for each indication encompass actions that can be monitored or validated using qualitative or quantitative measures. The ultimate TBL score is determined by the extent to which the initiative affects each measure of success.

1.5.1 How to Implement the Triple Bottom Line Decision-Making Tool

The TBL decision-making tool can be applied in three steps. The first step involves the company or organization's executives completing the TBL decision-making tool for a particular program. Consequently, users must utilize the drop-down menu to determine the effect of the initiative on each success measure (Table 1.2.). The decision-making tool will then create a score depending on the choices made. Executives can utilize the TBL decision-making tool to assess and analyze several delivery choices for the initiative, as well as to compare novel approaches with the standard business-as-usual (BAU) strategy. Alternatively, if they are collaborating with a group, they might opt to complete the TBL decision-making tool either collectively or individually and subsequently compare the outcomes, similarly to a procurement assessment.

Furthermore, additional suggestions on the stages of the TBL decision-making tool may include the following. For instance, if executives perceive themselves as incapable of addressing a performance metric but they believe that their effort will influence it to some extent, they must contemplate seeking assistance from another person or division that possesses expertise in that domain. Executives also have the option to choose "Unknown impacts." In addition, they have the option to offer concise explanations for the rationale behind their choices. These remarks can serve as justifications for executives' responses and will be valuable when completing the TBL Report Attachment Template for drafting a report for the Council/Committee.

Additionally, numerous efforts will not have the capability of significantly affecting or exerting influence over every metric of accomplishment. If there are any regions that are not covered by the initiative, mark them as "No impact/not applicable." In addition, executives may depend on familiar information wherever

Table 1.2 Options of the drop-down menu

Significant benefit	3 points
Positive impact	2 points
Meets minimum standards	1 point
No impact/not applicable	0 points
Unknown impacts	−1 point
Negative impact	−2 points
Significant adverse effect	−3 points

feasible, but certain TBL effects may be influenced by external circumstances and stakeholders beyond the company's divisions. Furthermore, it is crucial to discern these effects, whether they are detrimental or beneficial, since they may reveal possibilities to promote advancement in areas for which there is no complete accountability. Hence, executives have the option to utilize the comment box to provide clarification in cases where a success metric is impacted by other stakeholders or situations.

The next step involves identifying potential areas for enhancement in the initial TBL outcomes through modifications to the program. To do this, executives should adhere to the TBL Improvement Guidelines, which are as follows:

- At the very least, an effort should accomplish net sustainability benefits.
- Trade-offs between success measures are only acceptable if the whole project yields a net increase in sustainability.
- To enhance scores, it is advisable to minimize negative impacts and eradicate significant adverse effects, whether they are immediate or long-term. Additionally, it is beneficial to intensify the positive impacts of ongoing efforts by maximizing the benefits already being achieved. Furthermore, it is recommended to broaden the range of triple bottom line (TBL) success measures in order to generate more cobenefits by making improvements, regardless of their scale, in additional areas.
- Any conduct that has a substantial negative impact on present or future generations is only permissible if all other options are even more unfavorable.
- Any "Negative impact" or "Significant adverse effect" must be justified. It is also recommended for situations where the impacts are not known.
- Examine the materials provided in the Guiding Documents section and/or contact the individuals mentioned in the Contacts/Supports section for additional information and assistance.

Finally, executives should update the impact selections as necessary to reflect the changes made after adjusting or redesigning the initiative in accordance with the TBL Improvement Guidelines. This is the third phase. This will produce a new TBL score.

1.5.2 Limitations of the Triple Bottom Line Decision-Making Tool

The TBL decision-making tool does not offer an impartial sustainability grade or report card as it does not assess specific data or indicators. Instead, it serves as a tool for making decisions and encouraging modifications to our work methods. The Tool does not incorporate or assess the financial plan of an endeavor, nor does it analyze or assess the variations in cost among multiple alternatives. Hence, it is imperative to develop a comprehensive cost analysis or budget for your projects with a triple bottom line evaluation. Various elements contribute to determining the desirability and practicality of different options in the short, medium, and long term. These factors include legal consequences, engagement results, feasibility studies, technical reports, and others. The purpose of the TBL decision-making tool is to enhance, rather than substitute, the supplementary considerations that you, the Council, and leadership need to assess in order to make informed decisions.

Chapter 2
Multicriteria Decision Analysis: The Hierarchy of Criteria in the Triple Bottom Line Framework

2.1 The Process of Decision Analysis

The decision-maker's objectives and preferences are the primary factors in the development or selection of the most suitable alternatives. Decision analysis is a term frequently used to describe this procedure. Several alternative possibilities are assessed during the decision-making process. In such circumstances, it is not advisable to evaluate all feasible alternatives; rather, we should select the one that is most consistent with our objectives, preferences, lifestyle, values, and so forth [41]. To put it simply, decision analysis is the scientific examination of the process of making options. Examples of issues of choice include the optimization of reservoir operations, the development of flood protection options, and the selection of the most suitable technology for municipal water delivery. Goals, objectives, criteria, and attributes may be employed to characterize the preferences of a decision-maker (DM). Nevertheless, their interpretations are inconsistent. Goals are advantageous for establishing a precise and unambiguous level of success that one intends to achieve [42]. Objectives are pursuits that should be pursued to the fullest or generally represent the intended direction of development, whereas objectives are associated with desirable future performance results. Criteria are quantifiable and specific outcomes. The criteria frequently suggest the direction in which we should endeavor to progress. The objective of all decision-making challenges is to either achieve or prevent specific outcomes. One of the criteria should be the extent to which we achieve our objectives or prevent adverse outcomes. The alternatives are to attain, surpass, or not surpass [43]. The correlation between these three concepts is illustrated in Fig. 2.1. Among these characteristics are performance requirements, components, variables, features, and properties. Instead of "objectives" and "attributes," we will employ the term "criteria" in this volume as it is more frequently employed by DMs in the disciplines of environmental management and water resources.

K. Ragazou et al., *Triple Bottom Line and Multiple Criteria Decision Making Analysis*, Multiple Criteria Decision Making,
https://doi.org/10.1007/978-3-031-78045-5_2

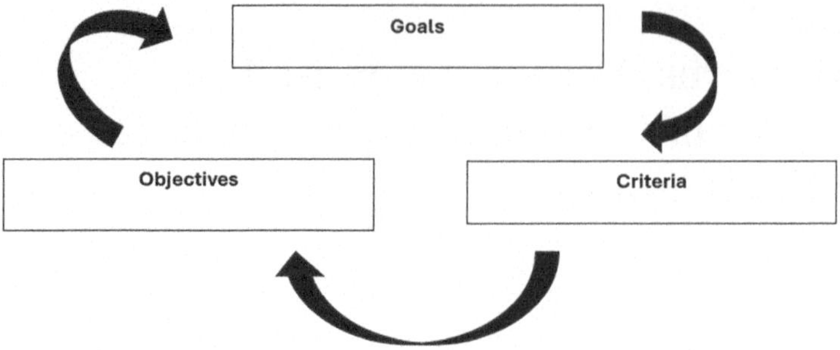

Fig. 2.1 The relationship between criteria, objectives, and purposes

2.2 What Constitutes the Problems Caused by MCDA

The decision-maker(s), criteria, and choices are the fundamental components of every multiple-criteria decision analysis (MCDA) problem. The classification of an MCDA problem is contingent upon the characteristics of these elements. The three components are explicitly defined as follows.

Individual(s) Responsible for Making Decisions The initial phase involves the identification of decision-makers (DMs). The decision-making process may involve an individual or a group of individuals or organizations in the context of a specific issue. There is only one DM in the first instance, while there are multiple DMs in the second condition. It is probable that the presence of numerous decision-makers (DMs) will result in the existence of unique preferences, objectives, criteria, and aims [44]. Consequently, it is exceedingly unlikely that any decision outcome would be equally satisfactory to all decision-makers. In such situations, it is essential to reach a collective resolution when the outcome is contingent upon the degree to which the different decision-makers take each other's interests into account. In essence, the outcome is contingent upon their willingness to collaborate. The problem may be classified as a multiple-criteria decision analysis (MCDA) problem if there are many individuals responsible for making judgments. The criteria for the current situation are those that are employed by a variety of decision-makers, as per Karamouz et al. (2003) [44–46]. A single-objective optimization problem is characterized by the presence of a single criterion and a single decision-maker. Linear programming, nonlinear programming, integer or mixed programming, dynamic optimization, or stochastic programming are among the methods that may be implemented in conjunction with the character of the issue. When conducting multiple-criteria decision analysis (MCDA), a single decision-maker may encounter complications when evaluating a multiplicity of criteria simultaneously. When a substantial number of decision-makers (DMs) are involved, a variant of criteria decision analysis (MCDA) can be employed to analyze the problem, as previously described. If the DMs have conflicting objectives and purposes, game theory may be a viable alternative. The most effective approach for addressing game theory scenarios that involve cooperative individuals is currently widely recognized as MCDA [45].

Alternative Options One is required to select from the following alternatives. Alternatives may be identified (i.e., discovered and examined) or even created (produced in the absence of a preexisting alternative). The decision space is the comprehensive collection of potential choices that can be assessed and implemented. A finite number of objects is frequently present in the choice space. A decision space is established with four available options when selecting a technology. In several additional cases, the decision alternatives are differentiated by continuous decision variables that specify the precise values at which the choice must be made. The capacity of the reservoir may vary within the range of the least and highest possible values.

Requirements The following characteristics or criteria must be present in differing degrees in each option. The criteria are assessed in relation to the degree to which the alternatives meet them. It is imperative to assess the quality of the choices available as they are limited to a predetermined set. The quality of any alternative may be determined by evaluating it in accordance with the criteria. Linguistic values, precise numbers, or stochastic or imprecise numbers may be used to differentiate these judgments. A criterion is deemed positive if the values are of a greater magnitude, suggesting that it is being assessed more favorably. In the same vein, a criterion is considered negative when a lower value suggests a more favorable assessment. MCDA issues can be categorized into two primary categories in terms of the alternatives.

It is imperative to evaluate the quality of the alternatives as we are required to choose one from a predetermined list of plausible alternatives. The magnificence of each option can be evaluated in relation to the criteria. Imprecise, random, linguistic, or unambiguous numbers may be employed to characterize these judgments. The numerical values that are higher suggest that a specific standard is more favorable. As a result, we classified it as a positive. A criterion is deemed negative when a lower number suggests a more favorable evaluation, as in the antecedent example. We have focused on two primary categories of MCDA challenges in relation to the numerous options that are available.

2.3 The Classification of the Criteria in the Hierarchy

Assessing the impacts of water resources and environmental initiatives in a linear fashion and depending on a single criterion to evaluate their benefits and make comparisons have become more difficult. However, water and environmental managers often assess, rank, and finally choose decision options solely based on a single criterion. A single criterion for selection can only be completely suitable under situations that are very clear and unequivocal. For example, they may choose the most extensive water transportation system from a lake, but they may have concerns about its cost-effectiveness, reliability, and environmental impact. This chapter will primarily concentrate on the dominating and crucial factors for water and environmental management within the context of sustainable development. Efforts are made to build and oversee water resource systems that are sustainable, meaning they are designed to maintain their ecological, environmental, and hydrological integrity. These systems attempt to fully support the objectives and goals of society, both now and in the future (ASCE 1998; UNESCO 1999). This chapter will provide a technique for choosing criteria that will be used to structure the hierarchy of the criterion. The technique will be shown by a comprehensive examination of a particular scenario. The following subsections will describe and assess diverse social, economic, and environmental concerns. Meeting these goals is essential for attaining the triple bottom line and guaranteeing long-term growth.

2.3.1 Environmental Criteria

Traditional water sources have experienced significant degradation because of their excessive utilization in recent years. The situation may deteriorate significantly in the future because of the impact of climate change and the escalating demands generated by a variety of factors. They will deteriorate the condition of the wetlands, groundwater reserves, lakes, and rivers. Every decision-making process should incorporate environmental considerations [47].

2.3.2 Social Criteria

The main goal of any water resource development plan is to enhance the social well-being of the area. Every attempt has a substantial impact on the well-being of society [48]. The following are the notable social components and externalities of these initiatives:

- Ensuring fairness is of utmost importance when allocating water resources. To mitigate social difficulties, it is crucial to regularly oversee and regulate both the upstream and downstream areas of a watershed. This equality is crucial for the success of a project as it will foster more public agreement and involvement.
- Traditional water rights holders are very diligent in protecting their rights. If a water plan violates a preexisting treaty, it will lead to significant conflicts within the community. Ensuring that the new water transfers effectively fulfill the requirements of these persons is crucial since their objection to the proposal would stem from the failure to do so.
- Projects that successfully reduce extreme poverty in the area and provide alternative job prospects are strongly preferred.

- The new environmental plans must align with the cultural and religious customs of the community.
- Occasionally, the construction of reservoirs, water transfer pipelines, or floodplain restoration projects necessitates the relocation of individuals from their homes, which can lead to a variety of mental and emotional stresses. As a result, initiatives that entail a reduced number of resettlements are more desirable.
- The construction of dams and their associated infrastructure has the potential to cause injury to historical structures, such as heritage buildings. It is essential to mitigate the extent of this harm.

Social characteristics are of great importance. However, the modeling of these entities is often challenging to complete since they are evaluated and ranked solely based on qualitative data, without any quantitative metrics included.

2.3.3 Economic Criteria

The primary criterion for evaluating various alternatives for water resource development is their economic feasibility. During the implementation of a project, it is possible to use effective techniques to evaluate the economic and financial results. For example, we may amplify the disparity between the total benefits of a project and its expenses. The foundation of our analysis is based on two essential economic principles: scarcity and substitution [49]. Humans have a natural tendency to make up for a lack of resources; thus, it is crucial to utilize them effectively. The differentiation between benefit and expenditure may be optionally replaced by the benefit-to-cost ratio. The importance is determined not by the exact amount of the benefit or cost but rather by the difference or ratio between the two. The internal rate of return is an additional metric used to assess the potential of investments. The breakeven interest rate is the rate at which the total expenses of an investment equal its profits.

2.4 Constructing the Hierarchical Framework of the Criteria

Various disciplines have previously been assigned a criteria hierarchy in the literature. Keeney and Raiffa (1993) state that any hierarchy of criteria must meet the following requirements [50]. The criteria require that it be functional, of the smallest possible scope, and capable of being deconstructed without any superfluous repetition. On the other hand, it is sometimes essential to compromise between these goals. For instance, the financial measure "benefit-cost ratio" may be calculated by considering the influence of the element "time of construction." Roy (1994) conducted research to investigate several options for choosing criteria, with the goal of facilitating a comprehensive review. Netto et al. (1996) carried out a case study in southern France using 13 criteria that were categorized into three primary groups: adaptability, dependability, and vulnerability. The process of establishing a hierarchy of criteria entails a repeated and progressive methodology. The process continues until stakeholders provide a set of criteria that has been unanimously agreed upon. It is essential to include a logical simplification of real-world situations in the hierarchy of criteria [51, 52].

2.4.1 The Importance of the Framework of the Value Management

Value management is committed to enhancing the capabilities of employees, motivating them, and fostering innovation in order to optimize the system's overall performance. Value management has advanced because of the integration of pragmatic principles and value concepts, which has resulted in improvements to existing methodologies. Miles is recognized as a pioneer in the discipline of value analysis during the 1940s and 1950s. To begin with, this approach was implemented with the objective of identifying and minimizing unnecessary expenditures [53]. The concept of value is derived from the connection between the costs associated with the execution of predetermined criteria in a model. The full fulfillment of requirements and the efficient use of limited resources are among the many factors that contribute to the increase in value. The consecutive phases of value management may be employed to organize the criteria [54].

Objective Identify the primary priority of the criterion. Function analysis is used to establish the major and secondary goals of the criterion, as well as the related correlations between the benefits and drawbacks. The primary functions refer to the main and fundamental aims, while the secondary functions are the unintentional outcomes that result from the implementation of the criteria. The main goal of implementing a feature like "Diversification of Financial Resources" is to reduce the budget of a government project. The secondary purpose of this feature is to establish a correlation between the project's financial needs and its clients, with the aim of reducing the economic risk associated with the project.

Idea Generation Employ ideation to establish new benchmarks that will increase value.

Assessment The new standards should be evaluated in accordance with their capacity to improve the overall value.

Action Plan Identify the techniques and tasks required to accomplish the intended objectives of value analysis and establish ongoing management structures to monitor project progress. An approach to integrated water resource management (IWRM) is presented in the subsequent section of the case study, which involves the evaluation and ranking of national water resource initiatives.

Detailed Reports and Statistical Analysis Generate the final report, which will encompass a summary of the procedure's results and a hierarchical arrangement of criteria.

Chapter 3
Environmental, Social, and Corporate Governance Factors Toward a Green Finance Framework

3.1 The Impact of Environmental Factors on Financial Stability and the Economy

The existence of global warming is supported by scientific evidence that is irrefutable. This will progressively and consistently have a significant impact on markets, society, and economies. The current climate change presents a significant threat to critical aspects of human existence, such as water supply, food agriculture, public health, and land and environmental management. Financial stability is significantly impacted by environmental concerns and climate change, as evidenced by an increasing volume of data. In order to address climate change and its associated hazards, it is imperative to implement legislative measures that are specifically designed to reduce global greenhouse gas emissions. By envisioning a future in which energy is obtained from alternative sources, a deadline is established for numerous investments that were previously without one. This may have a substantial effect on the evaluations. The consequences for individuals, corporations, and governments may be extensive as numerous critical sectors of the economy depend on current energy sources. Financial stability is endangered by the modification itself.

Nevertheless, the consequences of neglecting to alter the economy are considerably more severe. The combination of rising sea levels and soil degradation will result in a reduction in the number of inhabitable regions, which will pose a threat to urban centers. The safety of humans will be jeopardized by severe weather conditions. Values will be eliminated during this process. The financial system will be significantly affected by this situation. The recent catastrophic events serve as a stark reminder of the substantial influence that extreme weather changes can have on the economy. The 2010 heatwave in Moscow resulted in expenses that amounted to 1% of the Russian gross domestic product (GDP). In the same month, Pakistan experienced pervasive inundation that affected 20% of its territory, resulting in economic

K. Ragazou et al., *Triple Bottom Line and Multiple Criteria Decision Making Analysis*, Multiple Criteria Decision Making,
https://doi.org/10.1007/978-3-031-78045-5_3

losses of 5% of its gross domestic product (GDP). A year later, a significant industrial zone in Thailand was inundated, resulting in the destruction of manufacturing facilities owned by prominent Japanese industrial corporations. The country's gross domestic product (GDP) was reduced by 10% because of the damages. The National Oceanic and Atmospheric Administration (NOAA) in the United States has demonstrated that the frequency of catastrophic weather events resulting in damages exceeding $1 billion in the past 5 years has doubled in comparison to 1980. At present, there is an average of 11 such accidents each year. Businesses in numerous sectors, as well as numerous human endeavors, are being affected by the increasing frequency of these occurrences [55–57].

Typically, they are the result of substantial economic disruptions. The credit risk profile of institutions is significantly impacted by the extensive repercussions of this scenario, which in turn impacts the creditworthiness of debtors and projects for a variety of reasons. Everybody can observe the similarities between a financial crisis and a systemic environmental catastrophe. The investing community's majority frequently demonstrates a certain degree of predictability and rationality prior to a financial crisis as they are motivated by their own self-interest. They do this to optimize their own self-interest [57]. The cumulative effect of conceptually distinct functions has frequently jeopardized the overall well-being of the entire system, resulting in significant disruption in retrospect. Additionally, environmental hazards are predominantly concerned with protracted periods of time, whereas human financial behavior, which includes risk management, frequently prioritizes short- and medium-term perspectives.

The influence of environmental factors on insurance is readily apparent. Nevertheless, these attributes are of less importance in the financial sector when it comes to determining the most effective approach to loan provision and resource distribution. Financial institutions are required to incorporate climate-related factors and the potential systemic risk associated with climate change into their risk modeling and decision-making processes [58]. This is essential because it has the potential to significantly impact their financial stability and the rate of loan default. The process has commenced, and a framework is currently being developed to collect and mitigate these risks. The implementation of a standardized categorization system will improve comprehension and facilitate the development of more effective risk-reduction strategies. Currently, the legislation that governs the global environmental implications of corporations is restricted. It is imperative to establish environmentally beneficial rules and regulations for bank lending, trade, and investment operations to assess the risks associated with banks and accomplish the primary objectives of international financial institutions.

The primary objective of economic growth and financial development should be to achieve sustainability in the economic, social, and environmental domains. Mark Carney, the chair of the Financial Stability Board and governor of the Bank of England, has consistently emphasized the urgent necessity of prioritizing environmental considerations. The individual cautions that postponing action may result in a more challenging transition to a sustainable green economy, potentially having a substantial impact on financial stability. The 2006 study conducted by Lord Nicholas

Stern is significant in that it demonstrated that the cost of addressing climate change in its initial phases is significantly less than the cost of addressing its repercussions. Because of this, it is economically wise to address climate change and the financial repercussions that relate to altering climatic trends. Politicians have placed a greater focus on sustainable funding because of the Paris Climate Summit, which took place in December 2015, as well as the adoption of the 2030 Agenda for Sustainable Development. This is a trend that is quite promising. A comprehensive strategy for sustainable funding has been successfully implemented by the European Union (EU), which has garnered widespread recognition for its success. Both the High-Level Expert Group on Sustainable Finance and the Capital Markets Union Action Plan of the European Commission provide specifics about this plan. Both documents may be found on the European Commission's website. To hasten the transition toward a financial sector that is more ecologically and socially aware, European banks are keen to support and promote initiatives that have been developed by the European Union.

3.2 The Role of the Banking Sector in the Financing of the Transition to a Two-Degree Economy

Banks play a critical role in the financing of substantial changes in environmental policy. Due to their unique ability to facilitate the movement of money through activities such as lending, investing, and advisory services, this is the case. In addition, banks offer financing for innovative technologies that are crucial for the advancement of environmental development. Financing the transition to a low-carbon economy is the role of institutions [4, 59].

3.2.1 Transformation of Maturity

Although banks are becoming more active in the financing of sustainable activities, it is widely agreed that further efforts are necessary to attract private capital and to support the transition to a sustainable economy. One of the topics that is now being discussed by the public is the participation of banks in sustainable financing. There are some analysts who argue that banks have a limited role because of the disparity between the long-term nature of sustainable development endeavors and the short- to medium-term orientation of the banking system. As a result of this disparity, there are numerous analysts who acknowledge the significance of the banking industry in promoting sustainable development and suggest that banks should increase their involvement in financing sustainable projects. According to their argument, this would result in irregularities in the maturation of the financial statement and would make worries about the stability of the financial system even more severe. The debt capital markets are a good example of the mismatch that exists between sustainable investments and the financing solutions that are presently accessible. Although long-term investments are frequently required for environmental risk reduction operations, such as the management of air, water, and waste, and efforts to address climate change, such as the enhancement of energy efficiency, the credit duration for a significant portion of these activities is relatively short or medium term. This is because the credit duration is relatively short or medium-term. There is a typical lifetime for green bonds, which falls somewhere between the short term and the medium term [60–62].

Nevertheless, investors with extended time horizons, such as pension funds and insurance companies, endeavor to integrate environmentally sustainable assets into their investment portfolios by choosing financial instruments that are consistent with their long-term liabilities. The majority of initiatives that have the potential to benefit from green certification are typically of a modest to intermediate scale, with an average expenditure of 20 million euros. A minimal investment of 5 million euros is necessary for the undertaking. Furthermore, bond markets prioritize investments with the highest credit quality, as indicated by a public rating. Nevertheless, green initiatives that operate independently may receive a reduced credit rating. The banking industry must address a diverse array of risks to effectively manage supply and demand, including those that arise from counterparties that lack a comprehensive comprehension of the investment requirements required to manage their environment and climate change (ECC) risk profile [63]. Additionally, the industry is obligated to evaluate a wide range of potential hazards. Rather than substantial impediments to the procurement of long-term finance, financial institutions regard the processes of maturation transformation and risk assessment as essential obligations.

Rather, banks are obligated to construct their balance accounts in a restricted manner to:

- Ensure adherence to the current regulatory requirements.
- Ensure that capital is utilized efficiently within the current regulatory framework.
- Avoid the immobilization of money in anticipation of future changes in regulatory frameworks. This is a result of the volatile and unpredictable nature of the regulatory environment.

In addition, for financial institutions to participate in green finance activities and investments, a project must be fiscally, ecologically, and commercially sustainable. Although banks are unable to participate in the trading of sustainability outcomes, they are willing to consider both aspects. In spite of the fact that there is a certain degree of positive discourse regarding the promotion of long-term financing, it is not feasible to presume that the majority of short-term funding will be converted to "long-term." It is imperative to differentiate between "inherent" short-term finance, which necessitates the resolution of daily obligations or the fulfillment of immediate financial requirements, and "contrived" short-term finance, which is constrained by regulatory mandates, the complexities of evaluating long-term hazards, or the necessity of elevated risk and liquidity compensation [63–65]. A comprehension of the distinctions between the two categories of short-term finance is essential. The economic feasibility and financial feasibility of ventures are diminished because of these supplementary factors. The financial industry's long-term health must be improved by the implementation of specific regulatory or policy measures that resolve specific constraints. In addition, it is essential for financial institutions to develop business strategies and frameworks that are strategic and sustainable, as well as in compliance with the regulatory environment and government policies. These findings comprise a wide range of recommendations, each of which will be further elucidated upon in the consecutive sections [66].

3.2.2 Examples of Exemplary Methods

The banking industry acknowledges the significance of climate change and is dedicated to supporting EU policymakers and consumers in their efforts to achieve the objectives enumerated in the Climate Change Conference (COP) 21 in order to expedite the transition to a two-degree economy. Banks have substantial influence on the operating environment. In addition, they facilitate their clients' transition to more sustainable business models and contribute to this impact through their own financial portfolios and activities. Numerous institutions have implemented environmental mitigation strategies to lessen the environmental consequences of their operations and investments. However, their most significant contribution is the provision of financial solutions for organizations and initiatives that promote environmental sustainability [67, 68].

Numerous financial institutions have instituted ambitious objectives to simultaneously decrease their investments in carbon-related sectors and increase their investments in the renewable energy sector. Financial institutions are offering a wide range of environmentally responsible solutions and services that prioritize sustainability while concurrently fostering innovation and excellence. Renewable energies, energy conservation, and environmental preservation are financially supported by banking activities. These measures are applicable to a variety of entities, including public administration, retail consumers, small and medium-sized enterprises (SMEs), large corporations, start-ups, third-sector organizations, and significant domestic and international initiatives [69, 70].

3.2.2.1 Corporate Finance

Business finance is made accessible to corporate users to facilitate initiatives such as solar panels, biomass plants, hydropower plants, and energy efficiency measures. We are currently engaged in the development of agricultural protection initiatives to mitigate the extent of crop damage caused by hail, fire, and rain [71].

3.2.2.2 Retail Banking

Financial institutions provide loans to the retail sector for environmental initiatives. The program includes the installation of solar and photovoltaic panels, the energy-efficient renovation of dwellings, and the acquisition of eco-friendly equipment. In addition, financial institutions are actively involved in the European Mortgage Federation-European Covered Bond Council (EMF-ECBC) program, which specifically targets ecologically sustainable mortgages. Nevertheless, the main goal of the typical individual investor remains a blend of enhanced financial performance and decreased financial risk. Diversification sometimes involves choosing items that are seen as having increased risk or worse performance due to their environmental, social, and corporate governance (ESG) characteristics [72, 73]. The green bond and investment fund market is mostly influenced by major institutional investors. These investors give the most importance to issuers who demonstrate the highest level of creditworthiness, have moderate durations, and participate in particularly major public transactions. In addition, the investment product offering has not undergone significant changes because of the acknowledgment of the values that millennials express via their purchase choices.

3.2.2.3 Technology That Is Digital

Moreover, financial institutions are actively investigating the potential of using digital technology as a catalyst for advancing environmentally conscious finance in the areas of investments and loans, with the objective of increasing awareness. A notable innovation is the capacity of a bank's debit or credit card to monitor the environmental impact of the transactions conducted by cardholders. The correlation between ecological finance and financial technology, often referred to as FinTech, is still in its nascent phase. However, the process of converting the financial system into a digital format has resulted in a growing acknowledgment of the capabilities of financial technology (FinTech) in advancing sustainable development. Furthermore, this expansion includes artificial intelligence, mobile payments, and crowdsourcing, along with other technology. It is quite probable that the banking industry will see an increase in the establishment of partnerships and collaborations between established banks and emerging FinTech companies that provide novel products and services to the market [74, 75]. The proliferation of digital platforms and Fintech enterprises has significantly expedited the evolution of banks, increased the process of innovation, and facilitated the construction of new, mutually beneficial relationships. Operations and collaboration should occur under a legal framework that guarantees fairness for all parties offering comparable services. The "Green Digital Finance Alliance" platform exemplifies the collaboration in question. This platform was established with the purpose of tackling the potential of FinTech-driven business innovations to transform the financial system in a manner that is in line with the requirements of environmental sustainability. The Alliance comprises forward-thinking financial institutions dedicated to using digital technology to promote environmentally aware finance in the areas of lending, investing, and insurance. In addition, banks actively engage in funding initiatives that directly allocate their own capital to the energy, infrastructure, and environment sectors. This is done with the intention of improving the social and economic progress of local communities [76, 77].

3.2.2.4 Private Equity Firms

Within European banking groups, private equity (PE) companies are progressively using environmental, social, and governance (ESG) criteria, or ESG key performance indicators (KPIs), to oversee portfolio enterprises as a component of the investing process. Many private equity firms mandate that their portfolio companies provide reports on their nonfinancial key performance indicators (KPIs), which include the assessment of their carbon impact. In addition, private equity companies provide financial backing for ecological infrastructure projects, including biomass plants, hydroelectric or solar power plants, and waste treatment facilities [78, 79].

3.2.2.5 Impact Investment

The notion of impact investing provides financial institutions with an alternative avenue for engagement in solutions that have a significant impact on society or the environment while still generating a financial return. A few European institutions have established impact funds, which can exclusively invest in social organizations that have a positive impact on both society and the environment. Impact investing is a potential source of financing for initiatives that are less hazardous and have a reduced potential rate of return [80].

3.2.2.6 Green Revolving Credit Facilities

In recent years, there has been a rise in the number of financial institutions that have started providing green revolving credit facilities. These facilities are characterized by margins that depend on the borrower's ability to maintain their financial stability. Consequently, the creditor is given an extra economic motivation to improve the long-term sustainability of the relationship [81, 82].

3.2.3 Consideration of ECC Risk in the Regulatory Framework and Decision-Making Procedures of Financial Institutions

3.2.3.1 Integrating ECC Threats into Banks' Risk Management Systems and Strategies

The degree to which investors are susceptible to environmental concerns, including pollution, carbon emissions, and climate change, impacts their evaluation of credit risk. Neglecting material concerns may result in these challenges being perceived as a breach of the ethical principles that underpin society, as well as international accords and treaties. The absence of adequate comprehension poses a significant risk as it may result in the revocation of the operating license, penalties, legal proceedings against the management, a loss of market trust, consumer loss, and damage to reputation. It is crucial for the financier to comprehend and effectively manage this type of risk [83–85].

The concept of an indisputable "true value" may now be overly idealistic. It is imperative to evaluate techniques on a more comprehensive basis than merely financial performance indicators to achieve widespread adoption and use in the market. It is imperative to integrate ECC risk indicators, including the probability of defaults (PDs) and loss given defaults (LGDs), into credit ratings to differentiate between financially stable and financially unstable enterprises, as determined by the ECC screening criteria. It is essential to acknowledge, evaluate, and incorporate a company's environmental accomplishments into its overall performance assessment, including their monetary value. It is imperative to systematically integrate financial and ECC ratings to effectively resolve the issue rather than view them as separate entities. Additionally, it will require a significant quantity of time to compile a dataset that is sufficiently extensive.

Assessing the creditworthiness of institutions is crucial for determining their competitive standing. Consequently, banks' risk departments are expected to use standardized information sources, together with their own knowledge and skills, to evaluate environmental, social, and governance (ESG) risks. Financial organizations do audits of environmental due diligence to assess the environmental and social risks linked to investment projects [85–87]. The vetting process is continuously carried out over the whole period of these projects, and it is also included in their internal credit regulations and processes. Specialized committees and departments have been created to thoroughly examine and reduce environmental and social risks while also implementing suitable protocols for their staff members. In addition, a multitude of financial institutions have adopted climate risk management models that assess the risk linked to climate change at many levels, including sectors, subsectors, and individual businesses. These models have been designed to alleviate the impacts of climate change. These models evaluate the costs related to implementing policies aimed at reducing the hazards of climate change.

The Environmental and Social Management System Standard (ESMS) has been effectively integrated into the internal operations and procedures of numerous financial institutions. This was achieved through the establishment of strong partnerships with renowned global organizations, including the European Bank for Reconstruction and Development (EBRD) and the International Finance Corporation (IFC). The Enterprise Security Management System (ESMS), which is frequently abbreviated as ESMS, is a comprehensive collection of procedures that are seamlessly incorporated into the internal operations of financial institutions [86]. This organization ensures that the requirements and conditions for loans and other forms of credit follow the standards and laws established by the International Finance Corporation (IFC) and the European Bank for Reconstruction and Development

(EBRD). Both new and current sources of finance that financial institutions provide are pertinent to the statement. In other words, it is a highly effective approach to the identification, evaluation, management, prevention, and mitigation of environmental and social hazards that may arise because of the business operations of either the financial institution or the borrower [87, 88]. Nevertheless, financial institutions incorporate it into their evaluation of the credit risk they encounter as a result.

Implementing the Environmental Sustainability Management System (ESMS) results in a decrease in the expenses associated with environmentally friendly products and services in comparison to their traditional counterparts. As of 2017, Greek bank clients are required to submit their "Environmental Risk Assessment" paperwork. This tool is designed primarily to assist lending institutions in evaluating environmental and social concerns. If it is deemed necessary, a course of action, including corrective measures, must be provided according to the manner outlined in the ESMS literature. While some institutions have already integrated environmental concerns and dangers into their governance, strategy, operations, and risk management, it is acknowledged that more steps may be implemented at the European level. It is essential to have a thorough sustainability strategy that has clearly defined roles and accountability at the top levels of leadership [89–91]. Currently, the main objective is to measure and evaluate the environmental expenses and advantages of new projects utilizing analytical and modeling methods that are still in the first phases of development. The absence of a cohesive classification system, well-defined reporting procedures, and reliable data further adds complexity to the intricacy of risk assessments. The agreement on fundamental principles reinforces the concept of gathering and disseminating exemplary instances that may be presented to the public. The European Banking Federation is reaffirming its commitment to advancing optimal methodologies and enhancing awareness within the European banking industry.

3.2.3.2 Regulatory and Supervisory Frameworks That Account for ECC Risks

The existing regulatory structure in the banking industry may not be sufficiently prepared to effectively reduce systemic ECC risks. The lack of a universally acknowledged taxonomy and a standardized reporting mechanism is a major contributing element to this problem. We fully support the Basel Committee's clear acknowledgment of environmental difficulties and the growing influence these issues have on the stability and long-term viability of the economy. These difficulties are becoming more common and pose a substantial threat to institutions and the stability of the financial sector. Prior to the implementation of any Pillar I or Pillar II standards, it is necessary to develop a consistent taxonomy and disclosure structure. This is because it would be impractical to enforce any of these restrictions. Environmental risk assessment should be integrated into the examination of credit risk, operational risk, and, to a lesser degree, market risk. It is crucial to assess the degree to which environmental risk has been included in the appraisal of these other concerns [92, 93].

The regulatory and supervisory framework must give top priority to identifying and managing systemic risks. These risks include those that result from poor governance or weak ECC risk management. Furthermore, it should expressly take into account the possible risks connected with certain assets that, from an environmental standpoint, cannot be sustained for more than a few years. The annual Internal Capital Adequacy Assessment Process (ICAAP) is the recommended approach for documenting the management of ECC risk. Credit, market, and operational risk evaluations should be used to record hazards that are not viable. Banking authorities must engage with banks to provide new solutions for environmental challenges, as suggested by the presented information. To improve understanding of the banking industry and its ability to address future environmental concerns, it is crucial to collect the necessary data and conduct an analysis [93–95].

First Pillar

Basel III requires financial institutions to assess the impact of certain environmental and climate change risks on their operational and credit risk exposures. According to Basel III, specifically Pillar 1, Paragraph 510, financial institutions are required to efficiently handle the risk of environmental liability related to collateral, which includes the existence of dangerous chemicals on a property. Financial institutions are obligated to comply with this rule. Nevertheless, the overall attitude of the exposure is a significant aspect that is deemed risky in line with the Basel III Accord. Short-term corporate loans (1–3 years) have a lower risk weight on the balance sheet compared to long-term project finance loans (7 years or more) for non-publicly traded or off-balance sheet enterprises. The increased risk associated with the second category of loans is amplified by their longer duration and limited alternative options. As a result, the risk-weighting framework increases the capital requirements for banks that provide long-term financing for environmentally beneficial economic activities in project finance [96]. This is achieved by raising the capital requirements. This architectural style sometimes requires much higher financial inputs compared to other designs because of its longer lifetime.

Second Pillar

The Supervisory Review and Evaluation process (SREP) employs a rigorous approach to evaluate the additional risks that institutions encounter, in addition to the Pillar 1 procedure. This assessment is conducted to protect financial institutions. To ascertain the long-term sustainability of the bank's business strategy, it is necessary to evaluate whether it is concentrated on a few specialized areas or diversified throughout other industries. This is a very critical component that requires comprehensive examination. To guarantee the effectiveness of SREP, it is necessary to prioritize the bank's governance and include risk management as part of managerial responsibilities. The primary focus is on identifying solutions for the challenges associated with potential financial depletion and the risk of insufficient liquidity. Environmental risk considerations may be included into the supervisory framework and considered when revising the assessment methodology of the Basel Core Principles for Effective Bank Supervision. Banking supervisors are required to evaluate the use of hypothetical situations in their stress testing as part of the Supervisory Review under Pillar 2. These hypothetical scenarios will assess the financial implications of providing funds to businesses that are either ecologically sustainable or environmentally unsustainable. An assessment will be conducted throughout a certain timeframe to ascertain the effect on the stability of the financial state [97–99].

Given the importance of managing risk and maintaining competitiveness, it seems that the most pragmatic options in the foreseeable future would include implementing standardized economic capital calculation (ECC) scenarios and conducting stress testing. These measures are anticipated to result in various non-binding actions, including the allocation of resources toward knowledge development, the inclusion of ECC-related terms in credit agreements, the collection of data on ECC risk-related defaults in banks' portfolios, the exchange of interbank data, and the provision of financial support to ECC-related startups or research initiatives. It should be noted that these operations do not necessarily need additional capital or capital guidance measures.

It is advised to avoid using a Pillar 2 add-on, which is often suggested for "brown" threats or when institutions have not conducted adequate ECC risk assessments. Financial institutions aiming to include ECC (Environmental, Social, and Governance) considerations into their loan allocation rules or assess the risk associated with a particular borrower need supplementary data that is not fundamental to their core operations. They depend on obtaining information from other sources. The financial services industry should not be unjustly punished for the restricted accessibility of such information and data [99].

Furthermore, credit institutions conduct an evaluation of their internal capital adequacy (ICAAP), which is then evaluated by their regulatory authorities. This analysis enables them to identify significant risks and effectively convey their strategy for addressing them. This would include several elements, including but not limited to environmental hazards. Credit institutions evaluate their regulatory capital needs using a stress test exercise, which entails examining two scenarios: baseline and stress. This experiment investigates the potential risks that companies may face while transitioning to a low-carbon economy. The research considers several scenarios that assess the possible effect of this shift on the growth of the Gross Domestic Product (GDP).

Third Pillar

To ascertain the feasibility of banks disclosing information regarding their vulnerability to systemic risks and their strategies for mitigating them, bank regulators should assess Pillar 3 (Market Discipline). Consistently evaluating environmental hazards across multiple countries.

3.2.4 Strategies for Advancing Sustainable Financing

3.2.4.1 Consensus Taxonomy

Although bank credit is the primary source of financing in the EU economy, accounting for approximately two-thirds of investment, only a negligible proportion of this credit is explicitly designated as "green." This may be due to the absence of explicit and consistent definitions. Investors, companies, and financial institutions encounter a significant challenge in the form of a lack of precise definitions for green finance activities and products when seeking opportunities for investments that are environmentally benign. Among other things, these include green loans and green assets. Financial institutions have difficulty allocating financial resources for environmentally favorable initiatives and assets due to the absence of precise standards that serve as the foundation for internal budgeting, accounting, and performance assessment. It is imperative that all forms of reporting adhere to the principle of providing a viewpoint that is both objective and credible [63, 100]. The term "materiality" refers to the capacity to identify, quantify, and attribute a value to products and the hazards associated with them that are significant and pertinent. This concept does not offer any novel concepts. The concept of materiality already mandates that businesses and financial institutions incorporate climate hazards into their assessments of investment value.

Nevertheless, the definitions, measurements, and methodologies of the global financial industry vary from one business to another and from one sector to another. Convergence is necessary to effectively incorporate these hazards into the markets. The integration of climate-related risks into several sectors of the financial system is further complicated by the inherent dangers that are posed by the presence of various definitions and procedures. The Task Force on Climate-Related Financial Disclosure (TCFD) successfully achieved its objectives with the support of the Financial Stability Board. The Task Force has the responsibility of facilitating the use of financial measurement to advance the market and establish a universal international framework for green finance. A shared foundation for reporting, reviewing, certifying, and creating a protected EU green bond label could be established by setting up a reasonable level of harmonization in financial measurement, as well as common definitions and technical implementation measures or guidelines (after being appropriately translated into EU law) [77, 101, 102]. A common framework would be established to achieve this. In order to mitigate the risk of greenwashing, it would be essential to establish a standardized classification system that encompasses assets associated with carbon and green initiatives, as well as implement a set of minimal criteria. A company's exaggerated assertions regarding the environmental or social advantages of its products or services are referred to as "greenwashing" [103].

It is imperative to establish a taxonomy through a transparent process that involves numerous stakeholders. The potential for greenwashing has been a source of genuine concern, particularly in the context of emergent green asset classes such as index-linked green commodities, green-asset-backed securities, and environmental, social, and governance (ESG) funds. Green bonds offer the advantage of a more transparent structure; however, the absence of regulatory checks or penalties still poses a potential risk. There is a risk of excessive marketing of environmental issues in the case of other commodities due to the limited capacity of investors to appropriately assess investment strategies and the absence of limitations and penalties. It is imperative to have clarity when determining whether an investment can still be considered ecologically benign in the event of a change in scientific knowledge or data. Diesel engines were previously regarded as environmentally friendly; however, this perspective has substantially changed in the past decade.

3.2.4.2 Standardized Framework for Sharing Information

It is imperative to have analogous disclosures to address the insufficient information and internalize the costs associated with environmental and climate change performance, as well as the hazards associated with ECC. At present, there is a dearth of access to current data for the purpose of market research, effect evaluation, and other recommendations. The absence of data, inconsistent metrics (such as the absence of standardized energy labels for houses or variations in carbon dioxide (CO2) emissions of the energy mix across different countries), and the resulting lack of comparability present substantial obstacles to the development of new products that can be offered on a similar basis, the conducting of market and risk analyses, and the increase in awareness [104]. The process of data collection, key performance indicator (KPI) measurement, and quantitative data availability will be improved by the implementation of a European taxonomy and disclosure framework. Additionally, the implementation of disclosure regulations has the potential to increase the sustainability of lending portfolios. The European Commission should take action to establish a unified disclosure system that is widely supported on a global scale. This necessitates an acute awareness of the practical implications. Rather than endeavoring to establish its own definitions, the Commission should concentrate on utilizing the ongoing efforts of the FSB-sponsored Taskforce on Climate-Related Financial Disclosure, ensuring that it is applicable to the economic, legal, and regulatory framework of the EU. The significance and appropriateness of the information in relation to the situation, as well as the accessibility and expense of gathering information, are critical factors to consider. Institutions would be required to improve their internal capabilities and establish tools and procedures to implement a standard disclosure framework.

To establish appropriate situations and the corresponding analytical prerequisites, financial institutions must rely on the information provided by their counterparties. Additional effort is necessary. As a result, it may be more practicable soon to contemplate a qualitative disclosure system that is supplemented by voluntary quantitative information when it becomes accessible, such as through the ongoing efforts to implement the TCFD recommendations. The European Commission should implement a "comply or explain" framework that emphasizes transparency, accountability, and governance, rather than imposing excessively detailed regulations. This framework should be based on common conceptions. The implementation of a checkbox-style report is unlikely to improve sustainability and is more likely to increase the probability of engaging in deceptive practices. In addition, it is improbable that this would improve the understanding of risk and risk reduction as the reporting is contingent upon the independent evaluation of what is significant for the operations and stakeholders of the institutions [4, 59]. A term of 3–5 years would be suitable for the purpose of solidifying disclosures. It is imperative to establish widely agreed-upon taxonomies and comparable measurements on critical issues prior to implementing the necessary quantitative disclosures.

Furthermore, any reporting obligation should incorporate the materiality principle. It is imperative to establish and adhere to standardized definitions and techniques in order to prevent distortions and provide investors with accurate information. The TCFD's proposals have garnered the backing of over 100 multinational corporations, including numerous financial service organizations. We expect that a substantial number of these entities are currently making efforts to provide detailed information within the next year and are also preparing to undertake efforts to provide quantitative data.

In order to effectively address the practical difficulties of implementing the TCFD recommendations and incorporating them into the EU disclosure system, it is imperative to observe the UNEP Finance Initiative 8 with great care. It is worth considering whether Pillar 3 of the Basel Capital Accord, which incorporates the recommendations of the Task Force on Climate-Related Financial Disclosures (TCFD) and provides instructions for implementation from the European Banking Authority, could be a suitable method for banks to disclose information that is relevant to them. It is imperative to resolve the nonfinancial disclosure obligations of asset managers that are not contemplated by Basel III or Solvency II.

3.2.4.3 Environmentally Favorable Bonds Are Characterized by Consistent Criteria and Labeling

Projects and industries are financed by issuing bonds and securities on the market. Although the green bond market is growing in each reporting month, it still represents just about 1% of the overall bond market. While the requirements for green bonds are clear, their traceability remains complex. Nevertheless, the Climate Bond Initiative has performed an evaluation of the reporting of green bonds. A subsequent study has shown that 74% of green bonds issued before April 1, 2016, have integrated public reporting. The market's expansion may be enhanced by the adoption of stringent standards for green bonds, as seen in China. To create a consistent system across the EU, it is important for the EU to give priority to identifying possible possibilities for using or creating international standards and projects. The implementation of a uniform EU standard has the capacity to promote the expansion of the green bond market and prevent the formation of a separate market, as long as it does not impose overly strict requirements on those issuing the bonds. The European Business Federation (EBF) plans to commence the creation of environmentally beneficial financial instruments by deploying two specific kinds of bonds: the covered bond and the project bond. Furthermore, new bond classifications will be offered [105, 106]. A covered bond is a kind of financial product that uses mortgages as security. These mortgages are often of superior quality. The bond's price is mostly determined by the underlying securities, the mortgage institution's rating, and the nation's rating. We believe it is possible to create a green-covered bond by setting the necessary conditions for building new sustainable homes or renovating existing environmentally friendly houses. The local markets in these locations are using very effective methods. A green project bond is classed as "green" if the project it finances is deemed environmentally sustainable. Measures will be taken to secure the provision of second views, aiming to reduce the possibility of making misleading claims about the environmental sustainability of projects and dwellings. Independent parties will assess the risks connected with these perspectives, allowing investors to compare them with their own assessments. Additionally, a uniform system will be put in place to categorize, assess, and document environmentally friendly projects [27, 28].

3.2.4.4 Principles of Green Lending

A green financing policy usually includes favorable financial terms, such as lower loan rates, that banks provide to projects that are ecologically benign. It may also have limitations on projects that have a detrimental impact on the environment. Green lending involves many forms of financial support, including project finance, construction lending, and equipment leasing for businesses. It also includes personal home mortgage loans, motor vehicle loans, and green credit card services. The EBF is now engaged in the continuing endeavor to produce an industry-driven guideline for environmentally friendly financing as a member of the Global Green Finance Council, a coalition of financial sector trade groups [62, 107].

3.2.5 Collaboration Between the Public and Commercial Sectors with a Focus on Long-Term Planning and Coordination

Due to the substantial financial requirements, sustainable development initiatives should be financed through a variety of sources, such as private and governmental support. To optimize the utilization of public and private resources and mitigate hazards in the financing of sustainable development, it is essential to effectively implement public policy and finance. To expedite sustainable development and its funding, it is essential to ensure that governmental plans and policies are in accordance with the needs of the private sector, particularly the financial industry. The establishment of effective collaboration between the public and private sectors is imperative. The European Union (EU) must establish a comprehensive and enduring sustainability strategy, agenda, and route that will remain in effect until 2050 to provide market participants with greater predictability, transparency, and clarity. Additionally, this will guarantee that political objectives are consistent with the long-term sustainability finance initiatives [108]. One of the points of contention is the European Commission's desire to establish a legal system that encompasses the entire European Union or its preference for EU-level measures to operate in conjunction with the current policies of individual Member States. It is imperative that Member States are prepared to coordinate their national endeavors within a unified framework at the EU level in order to surmount specific obstacles within the European Union. It is imperative to improve the integration of financial and environmental policies, as well as the regulatory framework, across a spectrum of government agencies and departments. The entire process, from the development and execution of these policies to their enforcement, should be included in this coordination.

3.2.5.1 Risk Sharing in the Absence of a Market

Even though financial systems are capable of effectively managing risks, they frequently encounter challenges when it comes to tail-end risks and unanticipated calamities, which are occasionally referred to as "Black Swans." The fundamental essence of the system, which is composed of tiny individual enterprises, is the reason for this challenge. Subsequently, it is abundantly clear that engaging in risk sharing with public organizations serves a specific objective. However, we are cognizant of the numerous obstacles that this region faces. Concerns encompass a variety of various aspects, including the identification of the party responsible for the advancement of technology and the establishment of the regulatory framework for financial instruments and essential business processes. It will be crucial to establish a clear and concise understanding of the responsibilities and obligations that are to be transferred between private financial institutions and government organizations in the future. To guarantee that each euro spent by the government generates a substantial quantity of private investment, the objective of public financing should be to incentivize private and institutional investors to participate [109].

In order to preserve a consistent evaluation and market potential, it is essential to establish pricing and conduct a comprehensive evaluation of emerging and relatively new threats. In the context of this social endeavor, public investment institutions and public funds, such as the European Long-term Investment Fund (ELTIF), perform a function that is both substantial and distinctive. Despite the fact that the financial institution undertakes entire responsibility for risks that are expressly enumerated in the market, it is crucial for the public authority to emphasize its involvement in the resolution of risks that are not covered by the market. It is possible that the public body and the financial institution could collaborate to provide financial support for initiatives that would be unfeasible without the public's involvement

[110]. However, the necessity for these public institutions to implement measures to mitigate the threat they encounter is a cause for concern in this situation. The cost of risk is elevated because of these organizations' implicit guarantee, which ultimately leads to the exclusion of economically viable financial institutions from the market. The primary objective of these types of enterprises should be to identify uncharted market territories.

3.2.5.2 Government Grants Provided Directly to Individuals or Organizations

One potential alternative to consider is the provision of direct subsidies, such as tax advantages or subsidized financing conditions, that are analogous to those that are available in certain countries for the energy-efficient private house market. In Italy, the 2020 objective to decrease energy consumption is to improve the effectiveness of current initiatives and mechanisms, including tax deductions. In the United States, municipal bond owners are exempt from paying income tax on the interest they earn on these bonds, at both the federal and state levels. This exemption allows issuers to offer bondholders reduced yields. It is anticipated that the risk and price of financial assets will be influenced by the progressive elimination of unjustified subsidies to businesses that are heavily reliant on fossil fuels.

3.2.5.3 Monetary Policy

The accountability of national monetary authorities for the development of specific monetary policy measures, such as the acceptance of high-quality "green" assets from banks as collateral for central bank loans, the development of green finance guidelines, and the implementation of green reserve management strategies, should be assessed. These measures would enable institutions to allocate additional funds to environmentally sustainable economic initiatives.

3.2.5.4 Raising Consciousness

It is essential that banks and public authorities collaborate and coordinate their efforts to enhance the understanding of green finance among key stakeholders in order to address the insufficient communication between individuals seeking funding for their projects and financial contributors/investors seeking viable projects. At present, there is no explicitly established framework upon which this may be founded. The objective should be to integrate green finance as a critical component of the finance industry. The Global Green Council has implemented a variety of initiatives to facilitate and expedite the adoption of environmentally favorable financial practices. These initiatives encompass the establishment of a directory that offers information on European and global green finance policy initiatives, as well as the implementation of green lending principles. The directory's goal is to offer policymakers and participants in global and regional markets a reference guide that is both user-friendly and straightforward. This guide will provide information on the implementation of significant green, sustainability, and climate change initiatives by international and regional governments and industry bodies.

The European Investment Project Portal (EIPP) is a website that is publicly accessible and serves as a platform for the promotion of investment projects to potential investors worldwide. It is a critical element of the European Fund for Strategic Investments (EFSI). The EIPP's primary objective is to expedite and promote the advancement and realization of investment initiatives within the European Union. This is designed to make a substantial contribution to the expansion of the economy and the increase in employment rates. Considering the European Commission's proposal to extend the duration of the European Fund for Strategic Investments (EFSI), which is presently under discussion at the EU level, the European Investment Project Portal (EIPP) could be utilized to raise awareness of green finance. This would be in accordance with the EU's ambitious objectives, which were established at the COP21 Paris.

3.2.6 Growth of the Green Bond Market

The Monetary Authority of Singapore (MAS), the regulatory organization responsible for Singapore's financial sector, recently implemented the Global Green Bond Initiative, which is designed to encourage the issuance of green bonds in the current year. A maximum limit of 100,000 dollars per issuance is permitted for qualifying issuers to fully cover the expenses associated with procuring an external assessment of green bonds when they participate in the program. Additionally, market participants may exhibit a greater propensity to sustain these initial expenditures (or distribute them over a period of time) or provide equity as the risk-return characteristics of projects improve [107, 111].

This scenario would occur if the associated costs of CO2 emissions experienced a significant increase. It is appropriate to suggest the implementation of the "Structural Reform Support Programme" (SRSP) from 2017 to 2020, as outlined in Regulation (EU) 2017/82514. This is so because the information provided is reasonable. The primary objective of this program is to offer support to national authorities in the reform and enhancement of institutions, governance, public administration, and economic and social sectors in order to address economic and social challenges. In order to accomplish this objective, we will prioritize the implementation of structural reforms that foster development in the Member States of the European Union. These modifications will be structural, administrative, and institutional in nature. The objectives of enhancing efficiency, unity, competitiveness, sustainable development, job creation, and investment are being pursued through the implementation of this project. In the context of economic governance processes, this pertains to the provision of support for the efficient, transparent, and prosperous utilization of EU funds.

It is conceivable that this program could be advantageous to national authorities that are seeking to improve energy efficiency, diversify their energy sources, and implement climate change policies. Furthermore, it has the capacity to enhance the agricultural sector, promote the sustainable development of rural communities, and contribute to the expansion of fisheries. Furthermore, the European Investment Advisory Hub, a critical element of the European Fund for Strategic Investments, has the potential to be utilized in a more comprehensive capacity to provide improved technical assistance. It is conceivable that this may involve offering guidance to national authorities and project initiators to facilitate the identification, planning, and promotion of sustainable investment initiatives. This location has the potential to function as a consolidated technical consulting center for project financing within the European Union.

Additionally, the Carbon Disclosure Project (CDP) and other nongovernmental organizations (NGOs) may be able to aid investors in the decision-making process by establishing a global disclosure system that allows businesses to evaluate and regulate their environmental impact. Nongovernmental organizations (NGOs) may serve as the primary source of technical assistance for project financing within the European Union. Furthermore, their objective is to promote investment initiatives that prioritize environmental and social sustainability to the fullest extent. Nongovernmental organizations (NGOs) oversee a global system, including the Carbon Disclosure Project (CDP), which was previously known as the Carbon Disclosure Project. This system enables enterprises to assess and regulate their environmental impacts. Nongovernmental organizations (NGOs) may empower investors to make well-informed decisions.

3.2.7 Securing the Sustainable Feasibility of Lending Protocols in Export Credit Guarantees

The nations belonging to the Organisation for Economic Co-operation and Development (OECD) have made a commitment to not provide loans or insurance to governments and individuals in countries facing extreme poverty and debt. This assurance is applicable in certain circumstances, such as when the World Bank or the International Monetary Fund (IMF) neglects to carry out an assessment or when there is no connection with a grant component or official development assistance. The "sustainable lending guidelines" refer to the specific criteria established in the OECD Principles and Guidelines from 2008. These recommendations aim to encourage sustainable lending practices while providing official export credits to low-income countries. The objective of these recommendations is to ensure that lending procedures adhere to ecologically friendly standards. As severely indebted emerging nations choose to invest in green development instead of pursuing carbon-intensive options to satisfy their growing energy and infrastructure needs, they are missing out on a chance for economic growth. Moreover, the competitive environment is being disrupted due to the failure of certain member countries of the International Monetary Fund and/or the World Bank to comply with the regulations controlling sustainable funding. The European Business Federation (EBF) supports the suggestion made by the Business and Industry Advisory Committee (BIAC) to evaluate the criteria for sustainable financing in order to broaden the range of commercial loans linked to particular goods or activities. This may be conceivable for projects that fall within the scope of the 2014 Sector Understanding on Export Credits for Climate Change Mitigation and Adaptation, Renewable Energies, and Water Projects (CCSU), as well as for infrastructure investments.

3.2.8 The Regulatory Framework of Institutions in Relation to Green Finance

The absence of a comprehensive taxonomy and data is a hindrance to the advancement of green finance. Moreover, there is a dearth of intrinsic incentives to allocate funds toward the field of green finance. The absence of government incentives now impedes market players from directing capital toward ecologically friendly finance projects and the development of new green initiatives. This also impacts the accessibility of loans and funding for environmentally friendly initiatives, rendering them economically impractical. For instance, the price of green bonds and non-green bonds, both issued by the same corporation, is the same. There are no financial benefits to counterbalance the additional costs associated with the specific processes and reporting obligations for green bond issuance. Furthermore, the absence of monetary incentives fails to stimulate people to seek out new assets or projects that meet the criteria for refinancing via the issuance of a green bond. Implementing a strong incentive structure to promote higher lending toward projects that have a direct and beneficial influence on the environment, climate change, and low-risk industries will greatly enhance the efficacy of governmental environmental programs. This acknowledges the significant role that banks may have in conveying economic and political signals related to environmental issues. When determining incentives, it is crucial to consider the way the associated risk manifests and its effect on the financial system of the European Union.

3.2.8.1 Exemptions from Capital Requirements

There is evidence that suggests a direct and beneficial relationship between the financial success of investments that directly contribute to environmental benefits and their overall success. Hermes Investment's research suggests that organizations with superior environmental, social, and governance (ESG) ratings experience a decrease in credit default margins. The Industrial and Commercial Bank of China has reported that loans granted to environmentally friendly firms and sectors have lower default rates than those granted to non-environmentally friendly entities. Currently, there is a scarcity of information regarding the optimistic correlation between the financial performance of businesses and their reduced exposure to direct ECC risks. However, ongoing research is being conducted in this discipline. Despite the fact that green assets may imply a reduced level of risk compared to non-green assets, the financial burden of financing the energy transition remains a significant obstacle. In an economy where banks provide over 70% of the financing, it is imperative to persist in our efforts to acknowledge the advantageous nature of these assets for the energy transition in order to expedite and facilitate the funding of these assets. In the end, this will aid in the reduction of systemic risk by conducting a thorough analysis. Financing and investment in the energy transition, which is intended to mitigate climate-related risks and decarbonize bank balance sheets, would be encouraged by an appropriate prudential treatment that takes into account the positive environmental impact and reduced risk of these assets. Empirical research and regulatory considerations suggest that this will lead to a reduction in the capital necessary for direct financing and investment in these assets. It may also be applicable to low-risk industries that undergo ECC screening in order to provide them with assistance. In order to achieve this, it is essential to assess the feasibility of implementing a green supporting component, as outlined in Annex I. The financial risks associated with the initiatives and investments should be consistent with the prudential calibration for green and sustainable finance. The primary goal of macroprudential policy is to dependably guarantee financial stability. As a result, the capital necessary must be sufficient to effectively manage the realization of the associated hazards.

3.2.8.2 A Capital Treatment That Fluctuates According to the Asset's Time Horizon

Introducing a capital treatment for sustainable assets kept on the balance sheet would make it easier to finance their origination and future refinancing on the capital markets. The therapy would start with a small initial investment and gradually grow over time.

3.2.8.3 Green Finance with Less Short-, Medium-, and Long-Term Liquidity Restrictions

Liquidity challenges may arise because of the extended durations of long-term infrastructure investments that are designed to mitigate climate change and safeguard the environment. We propose that any promotional loan associated with green finance, such as pass-through loans, be considered an interconnected asset for both the liquidity coverage ratio (LCR) and the net stable funding ratio (NSFR), to resolve the current issue. This corresponds with the Commission's proposal under Article 428.f.1, as well as the provisions outlined in Article 424.6 of the CRR and Delegate Act 2015/61, as well as Article 26, Article 31 (9), and Article 33. Furthermore, we recommend that banks utilize the net stable funding ratio (NSFR) to evaluate the liquidity coverage ratio (LCR) and that they directly provide funding for environmentally favorable projects. This would assist in reconciling the disparity between long-term environmental objectives and the limitations imposed by prudential regulations on maturities' transformation. The net stable funding ratio (NSFR) may incorporate a component known as reduced needed stable funding (RSF) under specific circumstances to more precisely identify ecologically benign exposures. The potential for a review of this treatment may arise once a set of consistent criteria is established and additional data becomes available.

Chapter 4
ESG Factors Impact European Eco-Financial Institution Selection: A TOPSIS and Entropy Research

4.1 Sustainable Profitability and Environmental Responsibility in Entrepreneurship

Recent research has emphasized the need for financial institutions to implement environmentally conscious procedures. This study has shown the importance of implementing such methods. This shift has occurred due to several factors, such as heightened environmental consciousness among customers and stakeholders, a growing recognition of the long-term benefits of sustainable profitability, and stricter regulations imposed by the government [112]. Companies are shifting from just implementing green initiatives to fully integrating sustainability principles into their core strategy, therefore aligning themselves with the requirements of the global economy [113]. The concept of "ecopreneurship" is a crucial element of this transition since it is an essential component of this transformation. The involvement of financial institutions is very important due to their exceptional capacity to influence the actions of firms via the adoption of green finance and investment strategies. According to Noei et al. (2017), a major challenge in this industry is the absence of well-defined decision-making methods, particularly when there is a scarcity of historical data [114]. The ability of some techniques, such as entropy weight and the technique for order preference by similarity to the ideal solution (TOPSIS), to transform qualitative expert assessments into quantitative evaluations is fueling the increasing popularity of these approaches. This simplifies the arduous task of selecting and evaluating ecopreneurs based on the environmental, social, and governance (ESG) standards set in 2017 [115]. The study results indicate that it is important to provide personnel with environmental training and establish effective environmental management systems. These solutions enhance the long-term sustainability of financial institutions while also improving their performance in an environmentally conscious way [15, 116–118].

K. Ragazou et al., *Triple Bottom Line and Multiple Criteria Decision Making Analysis*, Multiple Criteria Decision Making,
https://doi.org/10.1007/978-3-031-78045-5_4

Currently, there is no universally accepted and approved definition for eco-entrepreneurship. However, Anand et al. (2021) and Johnson and Schaltegger (2020) provide noteworthy and current assessments of the existing body of research [119–121]. The United Nations' World Commission on Environment and Development (WCED) introduced the concept of "sustainable development" in 1987. The assumption is supported by recent research done by Cohen and Winn (2007), Dean and McMullen (2007), and Hall et al. (2010) [114, 122]. Sustainable development, as defined by the World Commission on Environmental Development (WCED), is the endeavor to satisfy the needs of society while also guaranteeing that future generations may meet their own needs without any constraints. Based on many study results, entrepreneurs may distinguish between eco-entrepreneurship and social-environmental entrepreneurship by considering the specific objectives they want to accomplish. Kraus et al. (2019) state that the primary goal of a sustainable enterprise (SE) is to achieve positive financial results, referred to as nonnegative nonfinancial returns, while minimizing any negative impacts on society or the environment. The primary objective of entrepreneurship in the environmental domain is to create positive results that are nonmonetary in nature. In addition, social entrepreneurship (SE) has mostly focused on environmental and social objectives, as opposed to the broader field of research on environmental, social, and governance (ESG) issues in management and economics. Sustainable development has primarily prioritized environmental and social goals. This has led to the government's objectives being assumed as a given. Dean and McMullen describe "eco-entrepreneurship" as the involvement of entrepreneurs in establishing an economy that is environmentally and socially sustainable. This research aims to provide a comprehensive definition of eco-entrepreneurship that incorporates all environmental, social, and governance (ESG) factors. More precisely, it recognizes the potential for eco-entrepreneurship to include elements that extend beyond ESG criteria. By doing this, we emphasize the significance of the dual-goal function. These initiatives are being implemented to tackle environmental, social, and governance issues while also generating favorable financial results and nonfinancial benefits [123].

Furthermore, there is a substantial amount of rapidly advancing research related to the entrepreneurial life cycle, including a wide range of perspectives. This study exemplifies the rapid progress of research. This body of work primarily focuses on the determinants of eco-entrepreneurship, the discovery and exploitation of opportunities for eco-entrepreneurship, and the consequences of taking advantage of these advances. In contrast, the scientific community has paid the least amount of attention to the results of eco-entrepreneurship. We highly recommend that the reader thoroughly examine the scholarly articles authored by Johnson and Schaltegger (2020) and Anand et al. (2021) in order to augment their comprehension [119–121]. The existing literature is fragmented, and there is a lack of research on the effects of software engineering in regard to our specific issue. The results of eco-entrepreneurship are impacted by programs that prioritize sustainability. The "triple bottom line" methodology is used to assess these accomplishments. This approach considers financial performance, societal influence, and environmental effects. Despite Anand et al. (2021) emphasizing the need to conduct a thorough investigation of the results of SE activity, this subject has only been the subject of a restricted number of research. This research focuses on two main subjects: (1) the effectiveness of funding and investing in green enterprises and (2) the impact of environmental, social, and governance (ESG) issues.

The "ESG effect" domain focuses on how eco-entrepreneurship contributes to environmental, social, and governance (ESG) objectives. The field of literature is greatly constrained by two major limitations. The optimal method for assessing sustainability is a subject of debate among scholars. The reason for this is that environmental, social, and governance (ESG) objectives are subjective, situational, and long-term and mutually reinforce each other. Furthermore, the establishment of a sustainable enterprise (SE) is inherently connected to opportunities for entrepreneurs. This is because the market lacks the ability to effectively remedy environmental damage. The results of the research conducted by Dean and McMullen (2007) and Cohen and Winn (2007) provide more support to this claim. Anand et al. (2021) principally emphasize that the majority of research on environmental, social, and governance (ESG) problems largely examines the effects on the environment [124–126].

Furthermore, our analysis is specifically focused on the field of eco-entrepreneurship and its investment performance, with a particular emphasis on the short term, as indicated by Bockel et al. [127]. The desire for financial profits by conventional players in the entrepreneurial financing market hinders the growth of self-sustaining firms. However, we are certain that crowdsourcing will successfully overcome this obstacle by supplying the required funds. According to Bockel et al., Block et al. (2018), and Vismara (2016), crowdfunding is regarded to be capable of providing the required funds to address this challenge [127]. Nevertheless, despite the enormous body of research conducted, there is a dearth of comprehensive investigation into the realm of social entrepreneurs and their mechanisms for delivering financial assistance. In their study, Cumming et al. (2016) discovered a notable association between the cost of hydrocarbons and the level of venture capital activity occurring within the clean technology sector of alternative energy. Cumming et al. (2017) reported that spotless technology enterprises had a higher success rate in reward-based crowdfunding campaigns on Indiegogo when they included a video pitch and were not for profit [128]. Hörisch's (2015) research revealed that there was no correlation between individuals' environmental awareness and the success of crowdfunding initiatives. This conclusion was reached by an examination of a subset of data collected from the same crowdfunding platform, which was similar to the original study done. The research conducted by Calic and Mosakowski (2016) revealed a positive correlation between the success of reward-based crowdfunding in technology, film, and video projects on Kickstarter and a focus on sustainability. In conclusion, Vismara (2019) showed that professional investors are less inclined to engage in equity-based crowdfunding projects that prioritize environmental protection. The existing literature on SE finance is still in its nascent phase. A comprehensive investigation has the potential to effectively address many significant deficiencies in the current knowledge base, particularly the "research gap related to the post-funding phase."

4.2 Drivers, Indicators, and Variables of ESG

Gaining a thorough comprehension of the procedures and rationales that dictate the impact of environmental, social, and governance (ESG) factors on a company's performance is very important. Based on performed research, it is recommended that experienced investors and asset managers prioritize the evaluation of environmental, social, and governance (ESG) regulations. This will help them optimize profits while also considering risk and enhancing risk management. According to surveys, 70% of investors make their investment choices based on sustainable investing criteria, while 14% of investors take it into consideration to some extent. Over 50% of asset managers and investors have embraced ESG principles to enhance their image and get long-term benefits. However, less than 30% of these persons engage in this practice with the intention of broadening their range of product offers. ESG principles have outpaced the growth of other corporate efforts and are expected to attract billions of dollars in future investments. Index providers have engaged in several discussions on their environmental, social, and governance values [128, 129].

Various environmental factors include sustainability, energy efficiency, pollution, carbon emissions, and the sustainable use of natural resources. Several significant social issues, including health, training, diversity, human and community rights, and privacy, contribute to workforce-related challenges. If a corporation fails to follow an environmentally acceptable policy, it may face legal repercussions and be subject to penalties [130–132]. Inadequate management of employees may result in subpar corporate governance, heightened rates of employee absence, diminished productivity, and disturbed customer relations, all of which are social phenomena. Inadequate management of staff may lead to decreased productivity as well. Furthermore, it might result in improper conduct regarding the upkeep of financial records, employee compensation, fraudulent activities, and the dissemination of erroneous information. Thomson Reuters employs nearly 400 distinct environmental, social, and governance (ESG) criteria, as documented in the literature [48, 129]. The firm provides the worldwide financial community with advanced indexes and related services, allowing investors to make informed decisions based on precise information. This index serves as an impartial benchmark for assessing the environmental, social, and governance (ESG) performance of a company. This is due to the growing significance of ethical investing in investors' decision-making processes. Investors want standardized reporting procedures since it is difficult to assess a company's investment potential and long-term viability using various measures. Therefore, investors need assurance that the firm is delivering precise information. Some metrics providers have developed techniques to report on environmental, social, and governance (ESG) factors because of this. Meanwhile, some metrics providers are enthusiastic about the creation of consistent standards [124, 125].

It is important to note that climate change has had significant impacts on the financial and economic development of enterprises. Consequently, there has been a rise in awareness, prompting investors to establish a connection between the management of climate change risks and economics. The growing emphasis on research into the environmental hazards associated with climate change illuminates the patterns that influence the commercial, financial, and economic sectors. Climate change is linked to several forms of physical dangers. Examples of these dangers include cyclones, wildfires, floods, and the adverse effects of interruptions in supply networks or financial markets. The significant impact of climate-related issues on financial assets is leading to an increasing probability of their occurrence. McBrien et al. (2021) state that enterprises striving to reduce their dependence on fossil fuels and mitigate other risks encounter a substantial obstacle in the form of nonperforming assets [133, 134].

There exists a significant correlation between innovation and the sustained profitability of enterprises. It includes a broad variety of activities, such as evaluating the environmental impact of products at every stage, overseeing supply chains, obtaining resources, and enhancing the overall resilience of the firm. Although the first impact of social media may seem little, its consistent earnings may significantly enhance the efficiency of products, facilitate employee retention, and foster customer loyalty. The notion of corporate social responsibility (CSR) is often linked to these objectives. However, the acquisition of social support is often the most challenging aspect of organizational investment. This discrepancy arises from the divergent viewpoints across different countries about the definition of a secure and sufficient benchmark [135]. The COVID-19 pandemic has shown the importance of social factors, such as performance and image, hence emphasizing their relevance. There has been much discussion in the corporate community over the effectiveness of participative management in addressing certain social issues [136–138].

The criteria presented in Table 4.1 collectively demonstrate the increasing recognition and emphasis that are being given to ethical, environmentally responsible, and socially responsible activities in the operations of businesses. The Emission Score is a metric that assesses a company's efforts to reduce environmental emissions in the production and operational processes. This metric underscores the growing importance of environmental responsibility in the conduct of corporations [139–141]. It is anticipated that credible sources, including Huijbregts et al. (2005) and Schulze et al. (2018), will provide comprehensive empirical or methodological research on the efficacy of mitigation strategies and the consequences of emissions. We anticipate this. The Policy Energy Efficiency Score is concerned with the execution of strategies and processes that are specifically designed to enhance energy efficiency. It serves as an indicator of the measures that organizations have implemented to promote the sustainable and efficient utilization of energy. Huijbregts et al. (2005) and Wang et al. (2015) predict that the majority of their publications will concentrate on the repercussions of these policies, legislative measures, and energy efficiency programs [73, 142].

The objectives are as follows. The Water Efficiency Score enables the assessment of objectives for the efficient utilization of water and underscores the importance of water conservation in the enterprise's long-term sustainability. Mugisha (2013) and Lozano and Soltani (2018) conducted research that provides valuable insights into

Table 4.1 The selected criteria

Criteria	Description
C1. Score for emissions	Emission category scores measure a company's dedication and effectiveness in decreasing environmental emissions during production and operations.
C2. Policy assessment for energy efficiency	This includes a variety of methods, tactics, and procedures to boost company energy efficiency. It refers to a structured method or system that optimizes energy efficiency and promotes continual improvement.
C3. Water efficiency goals and metrics	This debate will address company water consumption reduction goals, whether short-term or long-term.
C4. Corporate policy and ethics rating	Information on adherence to international norms of ethical conduct and honesty in business.
C5. Evaluation of policy on fair competition	This includes respecting other firms' patents, copyrights, and intellectual assets and avoiding anticompetitive behavior, pricing manipulation, and monopolistic techniques.
C6. Score for CSR sustainability reports	A company issues an extrafinancial report to detail its environmental and social impacts. If a corporation releases an extrafinancial report in a language other than its native, we shall verify it as "true" and remark. Web-based nonfinancial reporting updated yearly is acceptable. A thorough yearly report with sustainability measures provides accurate data. The annual report CSR part should be important and relevant. Comprehensive reports may include simply quantitative data and may be fewer than five pages.
C7. Human rights score	A company's efficacy in abiding by the fundamental principles of human rights is assessed by the human rights category score.
C8. Gender diversity grade for the board.	The percentage of women who are members of the board.
C9. Customer satisfaction score	The customer satisfaction rate that the organization has declared.
C10. Labor policy score	Strategies to prevent the company or its suppliers from discriminating against workers or recruiting minors under legal age limitations. The company or its suppliers avoid unfair employment practices or recruiting young workers who breach minimum working age regulations.

the water efficiency techniques and standards that can be implemented in corporate environments. The Policy Business Ethics Score is used to assess the organization's dedication to business ethics and integrity. This score emphasizes the importance of ethical conduct in the organization's operational procedures. Cho et al. (2021) and Schwartz et al. (2005) conducted a research study to examine ethical challenges, regulations, and compliance in the business sector [143]. The Policy Fair Competition Score assesses whether a company adheres to the principles of fair competition, in addition to emphasizing the importance of safeguarding intellectual property rights and promoting ethical competition. Encaoua (2002) and Fischbacher et al. (2009) conducted a study that examines the potential consequences of legislation, antitrust concerns, and competitive behavior. Both of these investigations may have substantial implications.

The Corporate Social Accountability Sustainability Reporting Score is a metric that assesses the extent to which nonfinancial information related to sustainability is disclosed. The score prioritizes the promotion of accountability and transparency in the discourse surrounding corporate social responsibility. The study conducted by Anas et al. (2015) and Baraibar-Diez & Odriozola (2019) focuses on the content, quality, and impact of sustainability reporting [126, 144]. The Human Rights Score underscores the company's responsibility to protect and advance human rights by assessing the company's adherence to human rights standards. Two academic works that evaluate the effectiveness of businesses' efforts to safeguard human rights through their actions are those of Arevadze (2021) and Cingranelli & Filippov (2018) [145].

The Board Gender Diversity Percent Score is a statistic that is employed to evaluate the proportion of female directors on a corporation's board. This metric underscores the importance of gender diversity in the management of corporate affairs. Yasser (2012) and Carter et al. (2003) are both involved in research initiatives that examine the current condition of gender diversity on corporate boards and its impact. The customer satisfaction score is a metric that demonstrates the importance of business strategies that prioritize the preferences and requirements of customers and measures the percentage of consumers who are satisfied with the company's products or services. Ittner and Larcker (1998) and Rosyid et al. (2018) both examine the methods employed to assess and improve customer satisfaction, as well as the outcomes of these methods. The Policy Labor Score highlights the necessity of combating unjust labor practices, including the exploitation of workers under the age of 18, as well as advocating for ethical employment practices and the preservation of workers' rights. Griffen (2018) and Triana et al. (2018) conducted a study that concentrated on ethical concerns, regulatory compliance, and occupational regulations.

4.3 Materials and Methods

The TOPSIS methodology is a systematic method that is implemented to investigate a variety of distinct components within a framework that has been effectively established. Entropy weighing is a method that entails the assignment of weights to a variety of variables or components in accordance with the degree of uncertainty or unpredictability, which is determined by entropy. The TOPSIS paradigm is a synthesis of the entropy and TOPSIS techniques. Kaur and her associates composed the citation in 2023. The entropy weight approach is the primary method for achieving the fundamental objective of this system, which is to determine the relevance of each evaluation indication. The subsequent determination of the evaluation item ranking is achieved by selecting the most optimal response. The TOPSIS technique is founded on the fundamental principle of identifying the optimal solution, which is the solution in which all attribute values have reached their optimal (or inferior) value in the alternative scheme. Dwivedi and Sharma (2023) are the purveyors of this information [36, 146, 147]. In order to ascertain whether or not an evaluation object is ideal, the proximity of the object to both the most favorable and least favorable solutions is assessed. An evaluation item is considered to be in the optimal position when it is located in close proximity to the most efficient solution and the furthest away from the choice that produces the least favorable results. Conversely, if it fails to meet these criteria, it is perceived as suboptimal. In an efficient fashion, the TOPSIS approach employs the information contained within the initial data, which is not subject to any specific constraints on the sample size. Atenidegbe and Mogaji (2023) and Zare Banadkouki (2023) emphasize that it provides the advantages of diverse functionality and minimal information loss.

Before presenting the approaches, we make the initial assumption that there are n alternatives $A = \{A_1, A_2, A_3, \ldots, A_n\}$ and m criteria $M = \{M_1, M_2, M_3, \ldots, M_m\}$, where $i \in A, j \in M, i = \{1, 2, 3, \ldots, n\}, j = \{1, 2, 3, \ldots, m\}$.

The matrix $X' = (x'_{ij})$nxm in Eq. 4.1 is a decision matrix of $n \times m$. The weights of criteria M can be represented by weight vector $W = \{w_1, w_2, w_3, \ldots, w_m\}$, which satisfy $\sum_{j=1}^{m} w_j = 1$.

$$X' = \begin{bmatrix} x'_{11} & x'_{12} & x'_{1m} \\ x'_{21} & \cdots & x'_{2m} \\ x'_{n1} & x'_{n2} & x'_{nm} \end{bmatrix} \tag{4.1}$$

4.3.1 The Entropy Weight Method

It is possible that the weight may be determined using the Shannon entropy weight technique by taking into consideration the degree of data dispersion (Saraiva, 2023). To standardize the $n \times m$ original decision matrix $\left\{x'_{ij}\right\}$., we begin by using the min-max technique from the beginning. Following that, we shift the standardized formula to the right by 0.001 units in order to simplify the logarithmic calculations that come after it.

$$x_{ij} = \frac{x'_{ij} - \min\left(x'_j\right)}{\max\left(x'_j\right) - \min\left(x'_j\right)} + 0.001, \text{where } i = 1, 2, 3, \ldots, n, \text{and } j = 1, 2, 3, \ldots, m. \tag{4.2}$$

Equation 4.4 has been employed to ascertain the entropy value, represented by the symbol e_j. Entropy evaluates the extent of data dispersion. A decrease in the entropy value is indicative of an increase in the degree of information content in the data, which occurs when the data exhibits a larger degree of dispersion. As the data becomes more concentrated, the entropy value will increase, indicating that the quantity of information contained within the data will decrease.

$$r_{ij} = \frac{x_{ij}}{\sum_{i=1}^{n} x_{ij}}, i = 1, 2, 3, \ldots, n, \text{and } j = 1, 2, 3, \ldots, m. \tag{4.3}$$

$$e_j = -\frac{1}{\ln n} \sum_{i=1}^{n} r_{ij}, i = 1, 2, 3, \ldots, n \text{ and } j = 1, 2, 3, \ldots, m. \tag{4.4}$$

The weight $w_j$is calculated using Eq. 4.5.

$$w_j = \frac{1 - e_j}{\sum_{j=1}^{m} (1 - e_j)} \tag{4.5}$$

4.3.2 The TOPSIS Model

The TOPSIS method, which was developed by Hwang and Yoon (1981), evaluates the options by evaluating the degree to which they are close to the best possible answers. For the purpose of determining the degree of proximity, the Euclidean distance between each objective option and the ideal and anti-ideal solutions is used instead. The suboptimal solution is described by the value that is the least favorable for each evaluation criterion, while the optimum solution is characterized by the value that is the most beneficial for each assessment criterion. In the end, the optimum solution is selected as the option that is preferred because of its dramatic contrast to the anti-ideal solution and its close similarity to the ideal solution. Each of the positive and negative criteria in the decision matrix in Eq. 4.1 is individually normalized at the beginning of the process. This is done in order to eliminate any differences in dimensions that may exist between the different criteria. It is of the utmost importance to emphasize the significance of adopting standardization across several dimensions by using the min-max technique shown in Eq. 4.6. In the process of creating the entropy weight, this technique makes it easier to strike a balance between beneficial outcomes and detrimental outcomes.

$$\text{positive}: x_{ij}^{+} = \frac{x'_{ij} - \min\left(x'_j\right)}{\max\left(x'_j\right) - \min\left(x'_j\right)}$$

$$\text{negative}: x_{ij}^{-} = \frac{\max\left(x'_j\right) - x'_{ij}}{\max\left(x'_j\right) - \min in\left(x'_j\right)} \tag{4.6}$$

$$\min\left(x'_j\right) = \left\{ \min_i x'_{ij} | 1 < i < n, 1 < j < m \right\}$$

$$\max\left(x'_j\right) = \left\{ \max_i x'_{ij} | 1 < i < n, 1 < j < m \right\}$$

Applying Eq. 4.6, the initial choice matrix is constructed. This matrix is then used in the construction of the dimensionless standardized decision matrix, which is denoted by infinity. The generation of this matrix is accomplished by normalizing the positive and negative criteria, as seen in Eq. 4.7.

$$X' = \begin{bmatrix} x_{11} & x_{12} & \cdots & x_{1m} \\ x_{21} & \cdots & \cdots & x_{2m} \\ \vdots & \cdots & \cdots & \vdots \\ x_{n1} & x_{n2} & \cdots & x_{nm} \end{bmatrix}, where\ i = 1,2,3,\ldots,n\ and\ j = 1,2,3,\ \ldots, m. \tag{4.7}$$

Furthermore, the decision matrix in Eq. 4.8 is derived by multiplying each element $v_{ij} = w_j \times x_{ij}$, where $w_j = (w_1, w_2, w_3, \ldots, w_m$ is obtained from Eq. 4.5 and meets the condition $\sum_{j=1}^{m} w_j = 1$ and x_{ij} is generated using Eq. 4.7.

$$V = \begin{bmatrix} v_{11} & v_{12} & \cdots & v_{1m} \\ v_{21} & \cdots & \cdots & v_{2m} \\ \vdots & \cdots & \cdots & \vdots \\ v_{n1} & v_{n2} & \cdots & v_{nm} \end{bmatrix} = \begin{bmatrix} w_1 x_{11} & w_2 x_{12} & \cdots & w_m x_{1m} \\ w_1 x_{21} & w_2 x_{22} & \cdots & w_m x_{2m} \\ \vdots & \cdots & \cdots & \vdots \\ w_1 x_{n1} & w_2 x_{n2} & \cdots & w_m x_{nm} \end{bmatrix} \tag{4.8}$$

According to Eq. 4.9, the positive ideal solution (PIS) is defined as the greatest possible value, and the negative ideal solution (NIS) is defined as the lowest possible value for each criterion. Both of these solutions are referred to as the ideal solutions. For the purpose of determining the distance that separates each option from the positive ideal solution (PIS) and the negative ideal solution (NIS), Eqs. 4.10 and 4.11 are applied.

$$PIS: P^{+} = \left\{ v_1^{+}, v_2^{+}, v_3^{+}, \ldots, v_m^{+} \right\} = \left\{ \left(\max_i v_{ij} | j \in M \right) \right\} \tag{4.9}$$

$$NIS : P^- = \{v_1^-, v_2^-, v_3^-, \ldots, v_m^-\} = \left\{\left(\min_i v_{ij} | j \in M\right)\right\}$$

$$d_i^+ = \sqrt{\sum_{j=1}^{m} \left(v_{ij} - v_j^+\right)^2}, i = 1, 2, 3, \ldots, n, \text{and } j = 1, 2, 3, \ldots, m. \tag{4.10}$$

$$d_i^- = \sqrt{\sum_{j=1}^{m} \left(v_{ij} - v_j^-\right)^2}, i = 1, 2, 3, \ldots, n, \text{and } j = 1, 2, 3, \ldots, m. \tag{4.11}$$

Ultimately, calculate the coefficient of relative closeness (RC).

$$RC_i = \frac{d_i^-}{d_i^- + d_i^+}, i = 1, 2, 3, \ldots, n \tag{4.12}$$

4.4 Results

The decision matrix that was produced over the course of the investigation for the research is comprised of three basic components, which are listed below. The matrix was built in order to facilitate the study. In the following sections, the phases and procedures that correspond to them are broken down into their respective components in further detail: the N1 technique is used during the first phase in order to standardize the data that is included within the decision matrix. This is done in order to ensure that the data is consistent. This approach entails translating the data into a scale that remains consistent throughout the whole process. This is done in order to make comparisons simpler to comprehend. This matrix was obtained following the completion of Step 1. Moreover, the last phase of the analysis included the process of assigning weights to each of the criterion *wij* and generating the weighted normalized matrix in accordance with the directions that were supplied in subsection 3.1 of the current study. This was done in order to guarantee that the findings of the study were correct and suitable for the situation. Throughout the course of this investigation, the entropy weight TOPSIS technique was used, which is a fact that should be kept in mind since it is very essential. The weights that have been assigned to each criterion are shown in Table 4.2, and they are arranged according to each fiscal year (FY).

As part of the scope of this investigation, information was gathered for 346 different financial institutions, which were situated in various parts of Europe. Furthermore, according to Refinitiv, a firm is considered to be ecologically responsible if it has a score that is greater than 0.75, and throughout this procedure, only 52 companies were selected for further study. In the process of assessing the eco-efficiency of businesses, the weights (Wi) that are shown in Table A1 in FY 2022 are used to

Table 4.2 Criteria weights are calculated using entropy weight TOPSIS

FY 2022										
Criterion	C1	C2	C3	C4	C5	C6	C7	C8	C9	C10
ej	−0.9942	−0.9849	−0.4925	−0.9899	−0.9332	−0.9990	−0.9976	−0.9889	−0.4888	−0.9898
$d = 1 - ej$	0.0058	0.0151	0.5075	0.0101	0.0668	0.0010	0.0024	0.0111	0.5112	0.0102
Wj	0.0051	0.0133	0.4445	0.0088	0.0585	0.0008	0.0021	0.0098	0.4477	0.0089

determine the relative significance of each criterion. These weights are generated by the entropy weight TOPSIS methodology. It has been determined that the criteria that are associated with customer satisfaction (C9) and water efficiency (C3) have been assigned the highest weights in this evaluation. This suggests that these factors are the most influential in this evaluation. According to the extensive emphasis that is now being put on customer-centric practices in corporate governance and the use of sustainable resources, this is in accordance with the current trend. On the other hand, the CSR Sustainability Reporting Score (C6) and the Human Rights Score (C7) are given much lower weights, which indicates that these aspects are less important in the overall assessment. This might be due to the fact that the processes of companies in specific sectors may have a higher degree of consistency or a lower degree of variance. The relatively low weights assigned to the remaining criteria in the context of the review process serve as proof of the significance of those criteria.

The association between service quality, customer happiness, and loyalty in a range of sectors has been the subject of a large number of research. These studies have provided useful insights into the elements that drive sustainable resource usage in corporate governance and customer-centric methods. According to Amin (2016), the relevance of service quality factors in the prediction of e-customer satisfaction and loyalty is emphasized. In a similar vein, Raza et al.'s research from 2020 highlights the significant influence that service quality plays in the growth of customer happiness and loyalty in the realm of online banking. These findings highlight the relevance of customer satisfaction as a vital consequence of service quality, which is compatible with the current focus on customer-centric methods in corporate governance. Such an emphasis is congruent with the significance of customer satisfaction. In addition, Andaleeb et al. (2016) provide direct insights into the direct influence that efficiency and customer satisfaction have on the profitability of businesses, notably in the banking industry. These studies provide credence to the idea that the achievement of operational efficiency and the happiness of customers are two factors that have an impact on the sustainability of company operations. In the context of water efficiency, the findings of the study that Lo and Id (2018) did on the water and wastewater service business in Italy shed light on the elements that determine the effectiveness of water utilities [148–150].

4.5 Conclusions

This study utilizes the concepts of entropy and the TOPSIS model to provide a thorough understanding of the process of identifying environmentally responsible financial institutions in Europe. The identification is based on their environmental, social, and governance (ESG) policies [115, 151]. The objective of this assessment is to provide a thorough comprehension of the procedure. Due to this revolutionary discovery, it has been acknowledged that providing environmental training to staff and implementing environmentally responsible management practices are crucial requirements for the functioning of these institutions[152, 153]. This method aims to illustrate how institutional tactics align with the concepts of sustainable finance. It does this by providing an elucidation of the correlation between the internal actions of a business and their broader environmental consequences. The findings of the study suggest that financial institutions that successfully use environmental management and training methods exhibit higher levels of eco-efficiency compared to those that do not adopt similar techniques. It is important to constantly remember that there is not necessarily a correlation between high TOPSIS and ESG ratings. The inference is that exceptional achievement in one field does not necessarily lead to significant success in another one, and this is a factor that should be considered [154, 155].

Based on the study findings, it is necessary to improve the strategic environmental, social, and governance framework, especially for businesses in lower ranks. The outcomes of this investigation have several consequences. Initially, it emphasizes the need for implementing strong environmental regulations and providing comprehensive training. Additionally, it delineates a precise course of action that financial institutions should follow to enhance their eco-efficiency. Both factors are crucial. Policymakers may use these data as evidence to formulate laws and incentives that encourage compliance with environmental, social, and governance (ESG) norms. These data might possibly serve as evidence for policymakers [124, 126, 156].

Additionally, it examines the correlation between the financial success of financial institutions and their capacity to ensure their future survival. The study effort used the entropy-TOPSIS model to assess the eco-efficiency of 20 different financial institutions located throughout Europe. The data clearly indicates a strong association between financial performance and sustainability, highlighting the significance of sustainable practices in achieving financial success.

Organizations that integrate environmental, social, and governance (ESG) factors into their business models often outperform their rivals that do not include these characteristics. Based on the results of this research, investors are progressively embracing sustainability standards to inform their judgments and choosing institutions that have robust environmental, social, and governance (ESG) attributes. The evidence indicates a growing ecological consciousness among individuals [134, 157–159].

This study provides empirical evidence to substantiate the integration of environmental, social, and governance (ESG) considerations into business models, which is an additional area of focus. This work also makes a valuable addition to the scholarly discourse on sustainable finance, which is another area of focus.

When doing more study, there is a vast array of potential options that may be explored. The proposal suggests expanding the scope to include a wider array of environmental, social, and governance (ESG) concerns, with the aim of conducting a more thorough assessment of the long-term sustainability of financial institutions. This would enable a more thorough evaluation of the institutions' ability to endure over an extended period of time. Longitudinal research is very beneficial in gaining useful insights into the long-term development of environmental, social, and governance policies. In order to have a thorough comprehension of the cultural and legislative elements that impact eco-efficiency, it would be beneficial to carry out comparative studies in various regions worldwide [18, 160, 161].

Studying the application of the entropy and TOPSIS model in different sectors might provide valuable insights into the broader trends and challenges related to the adoption of sustainable business practices worldwide. This may be achieved by examining the applicability of the model in various industries. Future studies may include further environmental, social, and governance (ESG) elements, alongside longitudinal analysis, to evaluate the effectiveness of ESG efforts over a longer prolonged duration.

Chapter 5
Controversies Surrounding Anticompetition and Anti-corruption Policies in Europe's Banking Sector: Assessing ESG Aspects Using the Entropy Weight TOPSIS Model

5.1 Anticompetition Concerns Arise as a Manifestation of Corruption Inside the European Banking Industry

The conceptualization and definition of corruption in both the public and commercial sectors have been the subject of a substantial quantity of research over the past few decades. The definition of the term is ambiguous as it may refer to a specific sector, a collection of autonomous enterprises, or an interdisciplinary perspective. The United Nations has designated corruption mitigation as a critical objective within the framework of the Sustainable Development Goals (SDGs) [47, 162, 163]. Corruption is a complex issue that impacts private companies, public organizations, and governments. The definition provided by the World Bank and Transparency International is highly influential and highly regarded. Corruption is defined by the World Bank as the exploitation of a public position for personal financial gain, while Transparency International (TI) defines it as the misuse of authority for personal gain. The public sphere is the primary manifestation of corruption, while the private sector is the secondary manifestation. The issue of corruption has been further exacerbated by the concealment of illicit funds by financial institutions and banks [164–166].

Embezzlement, misappropriation of funds, fabrication of financial statements and documents, cyber fraud, extortion, bribery, concealment of facts, and violation of law and regulatory frameworks are among the behaviors that characterize financial institution corruption. The financial industry is also involved in the transfer or reception of illicit funds, the violation of confidentiality, and the use of privileged knowledge for personal advantage. Furthermore, a significant portion of corruption in the financial sector can be attributed to the illicit activities of customers or organized criminal operations. This encompasses the utilization of fraudulent documents and the endeavor to intercept user data for the purpose of online banking.

K. Ragazou et al., *Triple Bottom Line and Multiple Criteria Decision Making Analysis*, Multiple Criteria Decision Making,
https://doi.org/10.1007/978-3-031-78045-5_5

However, the global financial crisis and the COVID-19 epidemic have reignited the interest of policymakers and scholars in the competition between banks, with a particular focus on allegations of anticompetitive behavior. These examples serve as evidence that corruption within the financial sector is increasing. It is a widely held belief in the academic community that an increase in anticompetitive activity could lead to financial instability. Additionally, there is apprehension that the international financial crisis and the pandemic-induced humanitarian crisis have resulted in an increased level of banking consolidation. Consequently, this could lead to the implementation of policies that are intended to prevent future instability and discriminatory competition because of the ethical concerns associated with financial institutions that are considered to be too large to fail [166].

The goal of European banks is to create a European financial system that is both efficient and effective in meeting the needs of European enterprises, investors, and citizens. It is essential to ensure that European firms, investors, and consumers have the requisite financial resources to prosper in a global economy that is exceedingly competitive, irrespective of their size, location, or nature. Despite its apparent simplicity, the establishment of a financial sector that seeks to reduce the occurrence of debts, high risks, and financial losses for both taxpayers and consumers has been a priority for European authorities and the financial industry over the past decade. The emphasis shifted from the efficacy or productivity of the financial sector to the safety, stability, and protection of investors in the aftermath of the severe global financial crisis, which threatened economic growth [167, 168].

The Scandinavian financial market is a prominent indicator of the severity of anticompetition concerns in the European banking sector. Scandinavia has consistently been acknowledged as one of the most progressive regions on the global platform [169]. The Scandinavian banks' involvement in a burgeoning scandal delving into anticompetitive behavior and money laundering is truly remarkable. Tchapchet Tchouto (2023) and Lindman et al. (2020) claim that these financial institutions facilitated the transfer of substantial amounts of illegal funds, totaling hundreds of billions of dollars, to offshore tax havens on behalf of Russian billionaires, corrupt politicians, and leaders of organized crime. For example, Danske Bank, a prominent financial institution in Denmark, was initially involved in illicit activities at its Estonian branch. Danske Bank became aware of the issues in 2017 when they were disclosed by Danish media [170, 171]. Danske Bank did not disclose the full extent of the transgression until September 2018, when the results of an inquiry conducted by an external law firm were publicly declared. In response to the increasing criticism, the bank acknowledged its "egregious failure to meet its obligation," and the chief executive, Thomas Borgen, resigned.

Additionally, the finance sector in Greece is a notable example as it is distinguished by high anticompetition scores. The Hellenic Competition Commission (HCC) is accountable for the enforcement of Greek and European Union antitrust legislation within Greece. This encompasses the implementation of Articles 1 and 2 of the Competition Act, as well as Articles 101 and 102 of the Treaty on the Functioning of the European Union (TFEU). Article 1 of the Competition Act and Article 101 TFEU prohibit any agreements, judgments, or activities that impede competition between enterprises. The exploitation of a dominant market position is prohibited by both Article 2 of the Competition Act and Article 102 TFEU. Financial institutions in Greece are closely linked to these anticompetitive behaviors. Misconduct allegations are frequently encountered by commercial institutions. Nevertheless, a number of prominent financial institutions have implemented low-cost loans as a strategy to secure and attract profitable operations that were previously conducted by investment banks. These allegations are frequently observed in the financial sector of Greece and other countries. Loan decisions were prohibited from being incorporated with other banking activities for an extended period. Any astute banker would vigorously deny any allegations that their financial institution was engaging in such conduct with the assistance of their legal counsel. The concept of connecting grants and other activities may seem simple; however, the actual demonstration of this correlation is often insufficient, and the constraints can be easily surmounted. For example, a financial institution can thoroughly evaluate all of its interactions with a client in order to ascertain whether to continue to provide financial assistance. Consumers are permitted to indicate their desire to form a business partnership with a bank in the interim if they are granted access to affordable financing. Conversely, the Romanian financial sector serves as an exemplary illustration of equitable competition in the financial market. According to the Competition Council, Romanian law prohibits any agreements or coordinated actions between financial institutions that impede, restrict, or distort competition in the Romanian financial sector. This encompasses the determination of pricing for the purpose of purchasing or selling, as well as any other conditions associated with trade [172, 173].

5.2 ESG Controversies: The European Finance Sector's Influence and Corruption in ESG Standards

Despite the endeavors, corruption persists of national and international regulatory entities, such as the International Monetary Fund (IMF), central banks, regulatory agencies, and adherence to Organisation for Economic Co-operation and Development (OECD) norms. Corruption directly contributes to the prevalent presence of poverty, greed, unemployment, weak institutions, and insufficient law enforcement. According to the Global Witness Organization (GWO), financial institutions are consistently used by corrupt persons, such as businesses, government officials, warlords, and criminals, to hide and validate their unlawfully obtained cash. Therefore, banks and other financial organizations are accountable for the hiding of illegal money. The bankruptcy of financial organizations in past cases, such as Bear Stearns and Lehman Brothers, may have been caused by the incompetence of bank executives, the participation of firms in illegal activities, and the inadequate corporate governance of banks. Regulatory authorities and academic scholars have undertaken inquiries and studies on the misconduct in financial institutions in reaction to instances of bank scandals.

In recent years, there has been an increased focus from policymakers, practitioners, stakeholders, and academics on the topic of corruption within the financial sector. However, the main emphasis is on the economic and financial variables that contribute to the formation and spread of indicators related to environmental, social, and governance (ESG) controversies rather than those that contribute to the incidence of extortion, fraud, and corruption. The word "ESG" encompasses the three key components of environmental, social, and governance, which go beyond economic and financial factors. The importance of these factors is growing in the realm of corporate management and financial investments (Agnese et al., 2023; Cernohorska, 2015; Khan et al., 2022).

The ESG controversies rating is a quantitative measure that assesses a company's susceptibility to environmental, social, and governance difficulties, as well as negative occurrences, as documented in the worldwide media (Agnese et al., 2023; Anita et al., 2023; Bang, Ryu, & Yu, 2023). Several scholars have lately performed studies on the significance of the environmental, social, and governance (ESG) discourse in connection to nonfinancial companies. Li et al. (2019) examine a dataset of 2613 enterprises from 39 countries throughout the period of 2009–2013. The authors propose that when faced with a disagreement, corporations will want to minimize the lasting negative impact on their credibility, reputation, and relationships with stakeholders by participating in symbolic corporate social responsibility (CSR) efforts. Aouadi and Marsat (2018) show a positive relationship between the increased market value of corporations and the discussion around environmental, social, and governance (ESG) problems. This is accomplished by examining a comprehensive dataset that encompasses over 4000 firms from 58 countries spanning the years 2002–2011. DasGupta (2022) examines a comprehensive dataset including 27 countries and covering the period from 2010 to 2019. This analysis is conducted in conjunction with a correlated research endeavor. According to the author, firms that are dealing with financial problems and are facing a lot of controversies related to environmental, social, and governance (ESG) concerns are more likely to attain a strong degree of ESG performance. In contrast, Marsat et al. (2022) highlight the significance of environmental disputes [48, 128, 129]. The authors provide evidence that the ability of a company to withstand and recover from challenges is significantly improved by its environmental performance, specifically in the case of an environmental dispute. They analyzed a dataset of 233 environmental disputes that occurred globally from 2010 to 2016.

When discussing corruption in the financial sector, policymakers, practitioners, stakeholders, and researchers usually focus on the economic and financial aspects that might lead to a rise in extortion, fraud, and corruption. They often neglect to take into account the elements associated with the rise and magnitude of environmental, social, and governance (ESG) problems. ESG, an acronym for environmental, social, and governance aspects, encompasses the three main components that go beyond financial and economic issues [125, 126, 156]. These characteristics are growing in importance in the areas of company administration and financial endeavors. The

ESG controversies rating evaluates a company's vulnerability to environmental, social, and governance difficulties and negative events as they are portrayed in the worldwide media. Several scholars have lately performed research on the impact of the ESG discourse on nonfinancial firms. However, there has been no clear investigation of the impact of ESG elements, namely, those that create problems and lead to wrongdoing in the financial industry.

Nevertheless, this study represents the first attempt to examine environmental, social, and governance (ESG) problems specifically inside the banking industry. The objective of our study is to assess the extent to which the dynamics of ESG discussions give rise to concerns about anticompetition, corruption, and illegal actions inside European financial institutions. To do this, we will analyze a panel dataset, including 344 European financial institutions spanning the years 2018–2022. We analyzed 14 criteria that are relevant to ESG controversies and are based on the three subpillars of ESG (Table 5.1) to evaluate the influence of the chosen ESG controversies components. We adhere to the categories and subdivisions defined by the reference data source.

5.3 Materials and Methods

The technique for order of preference by similarity to the ideal solution (TOPSIS) is a methodical approach that is used to develop evaluations by taking into account a variety of factors within a well-defined framework. Entropy weighing is the process of allocating weights to various components or variables based on the degree of uncertainty or unpredictability, as quantified by entropy. The TOPSIS model is a hybrid approach that integrates the entropy methodology with the TOPSIS method [188, 189]. The entropy weight approach is the primary objective of this system, which is to determine the significance of each assessment indicator. It employs a methodology to ascertain the hierarchy of evaluation elements by attaining the most optimal solution. The entropy weight TOPSIS approach is predicated on the principle of identifying the optimal solution, which is the one in which all attribute values have reached their highest (or lowest) value in the alternative scheme. This information is derived from the publication by Dwivedi and Sharma, which was published in 2023. When an evaluation object is the furthest from the worst option and the closest to the ideal solution, as determined by comparing their relative distances, it is considered optimal. In contrast, it is deemed suboptimal if it lacks these attributes. Entropy weight is effectively utilized by the TOPSIS technique, which does not impose any specific restrictions on the sample size. This strategy provides a minimal amount of information loss and a versatile operation [146, 190].

Prior to presenting the approaches, we make the initial assumption that there are n alternatives $A = \{ A_1, A_2, A_3, \ldots, A_n\}$ and m criteria $M = \{ M_1, M_2, M_3, \ldots, M_m\}$, where $i \in A, j \in M$, $i = \{1, 2, 3, \ldots, n\}, j = \{1, 2, 3, \ldots, m\}$.

Table 5.1 Criteria and their components of calculation

Criteria	Description	Measurement
C_1	Policy fair competition score	Does the corporation explicitly state in its code of conduct that it endeavors to be a fair competitor? – Encompasses adherence to the patents, copyrights, and intellectual properties of other companies, as well as avoidance of anticompetitive behavior, price fixing, and other monopolistic practices. – Extract data from the code of conduct section in each given report.
C2	Policy bribery and corruption score	Does the company explicitly state in its code of conduct that it endeavors to prevent bribery and corruption in all of its operations? – The company has implemented a policy in its code of conduct to address and prevent bribery and corruption in its operations. – Any report should take into account the information provided in the code of conduct section. – Legal compliance data is not taken into consideration. – This policy covers activities such as inappropriate or improper payments, special favors, extortion, or kickbacks [117].
C3	Bribery, corruption and fraud controversies score	Is the corporation now facing public scrutiny due to any controversies related to bribery, corruption, political donations, unethical lobbying, money laundering, unauthorized imports, or tax evasion [166]?
C4	ESG controversies score	The ESG controversies category score quantifies a company's level of involvement in environmental, social, and governance disputes and unfavorable occurrences as reported in international media [86, 115].
C5	Wages working condition controversies score	Is the company currently facing public scrutiny owing to controversies involving its employees, contractors, or suppliers, specifically related to wage disputes, layoffs, or working conditions [174]?
C6	Consumer complaints controversies score	Is the company receiving significant media attention due to consumer complaints or dissatisfaction specifically related to its products or services [175]?
C7	Product quality controversies score	Is the corporation currently facing public scrutiny due to an issue about the quality and responsibility of its products or services [176]?
C8	Accounting controversies score	Is the corporation currently facing public scrutiny due to a controversy related to aggressive or nontransparent accounting practices [173]?
C9	Responsible marketing controversies score	Is the corporation currently facing public scrutiny due to an issue related to its marketing methods, including the overpromotion of unhealthy food to susceptible consumers (Γθο ετ αλ., 2022; ΄Κασημιρι ετ αλ., 2019; ΄˜θ ετ αλ., 2023α)?

(continued)

Table 5.1 (continued)

Criteria	Description	Measurement
C10	Executive compensation controversies score	Is the corporation attracting public attention due to a scandal around excessive compensation for high-ranking executives or board members [80, 165]?
C11	Insider dealings controversies score	Is the firm currently facing public scrutiny due to a problem involving insider trading and other forms of share price manipulation [179–181]?
C12	Policy business ethics score	Does the organization explicitly state in its code of conduct that it is committed to upholding the utmost standards of overall business ethics? – Details regarding adherence to universal principles of corporate ethics or integrity. – Data derived from the part pertaining to the code of conduct [182, 183].
C13	Environmental controversies	Is the corporation currently receiving significant media attention due to an issue around the environmental impact of its operations on natural resources or local communities [184]?
C14	Emissions score	The emission category score quantifies a company's dedication and efficacy in mitigating environmental emissions during its manufacturing and operational activities [185–187].

The matrix $X' = (x'_{ij})\ n \times m$ in Eq. 5.1 is a decision matrix of $n \times m$. The weights of criteria M can be represented by weight vector $W = \{w_1, w_2, w_3, \ldots, w_m\}$, which satisfy $\sum_{j=1}^{m} w_j = 1$.

$$X' = \begin{bmatrix} x'_{11} & x'_{12} & x'_{1m} \\ x'_{21} & \cdots & x'_{2m} \\ x'_{n1} & x'_{n2} & x'_{nm} \end{bmatrix} \tag{5.1}$$

5.3.1 *Entropy Weight Method*

Shannon's entropy weight method can ascertain the weight based on the level of data dispersion [191, 192]. Initially, we employ the min-max technique to normalize the $n \times m$ original decision matrix $\{x'_{ij}\}$. Subsequently, we shift the standardized formula to the right by 0.001 units to simplify subsequent logarithmic computations.

$$x_{ij} = \frac{x'_{ij} - \min\left(x'_j\right)}{\max\left(x'_j\right) - \min\left(x'_j\right)} + 0.001, \textit{where } i = 1, 2, 3, \ldots, n, \text{and } j = 1, 2, 3, \ldots, m. \tag{5.2}$$

The entropy value, which is denoted as e_j, has been computed by using Eq. 5.4. Entropy quantifies the level of data dispersion. As the data becomes more spread out, the entropy number decreases, indicating that the data contains a greater amount of information. As the data becomes more concentrated, the entropy value increases, indicating a decrease in the amount of information contained in the data.

$$r_{ij} = \frac{x_{ij}}{\sum_{i=1}^{n} x_{ij}}, i = 1, 2, 3, \ldots, n, \text{and } j = 1, 2, 3, \ldots, m. \tag{5.3}$$

$$e_j = -\frac{1}{\ln n} \sum_{i=1}^{n} r_{ij}, i = 1, 2, 3, \ldots, n \text{ and } j = 1, 2, 3, \ldots, m. \tag{5.4}$$

The weight $w_j$is calculated by using the Eq. 5.5.

$$w_j = \frac{1 - e_j}{\sum_{j=1}^{m} (1 - e_j)} \tag{5.5}$$

5.3.2 TOPSIS Method

The TOPSIS approach, developed by Hwang and Yoon in 1981, evaluates the alternatives by measuring their closeness to the best solutions [43, 193]. The Euclidean distance is used to compute the proximity between each target option and the ideal and anti-ideal solutions. The ideal solution is determined by selecting the most beneficial value for each assessment criterion, whereas the worst prospective alternative is characterized by the least favorable value for each evaluation criterion. The optimum solution is selected based on its near-to-the-ideal solution and large departure from the anti-ideal solution, making it the most suited choice. Firstly, the positive and negative criteria of the initial choice matrix in Eq. 5.1 are individually normalized to eliminate any differences in dimensions among the different criteria. Emphasizing the significance of using the min-max approach in Eq. 5.6 is crucial for maintaining consistency in various domains. This technique simplifies the process of balancing the pros and downsides in the formulation of the entropy weight TOPSIS [36, 146, 147].

$$\text{positive}: x_{ij}^{+} = \frac{x'_{ij} - \min\left(x'_j\right)}{\max\left(x'_j\right) - \min\left(x'_j\right)}$$

$$\text{negative}: x_{ij}^{-} = \frac{\max\left(x'_j\right) - x'_{ij}}{\max\left(x'_j\right) - \min in\left(x'_j\right)} \tag{5.6}$$

$$\min\left(x'_j\right) = \left\{ \min_i x'_{ij} | 1 < i < n, 1 < j < m \right\}$$

$$\max\left(x'_j\right) = \left\{ \max_i x'_{ij} | 1 < i < n, 1 < j < m \right\}$$

Then the dimensionless standardized decision matrix x_{ij} is made by using normalized positive and negative criteria to make the initial choice matrix in Eq. 5.6, as shown in Eq. 5.7.

$$X' = \begin{bmatrix} x_{11} & x_{12} & \cdots & x_{1m} \\ x_{21} & \cdots & \cdots & x_{2m} \\ \vdots & \cdots & \cdots & \vdots \\ x_{n1} & x_{n2} & \cdots & x_{nm} \end{bmatrix}, where\ i = 1,2,3, \ldots, n\ and\ j = 1,2,3, \ldots, m. \tag{5.7}$$

Furthermore, the decision matrix in Eq. 5.8 is derived by multiplying each element $v_{ij} = w_j \times x_{ij}$, where $w_j = (w_1, w_2, w_3, \ldots, w_m)$ is obtained from Eq. 5.5 and meets the condition $\sum_{j=1}^{m} w_j = 1$ and x_{ij} is generated using Eq. 7.

$$V = \begin{bmatrix} v_{11} & v_{12} & \cdots & v_{1m} \\ v_{21} & \cdots & \cdots & v_{2m} \\ \vdots & \cdots & \cdots & \vdots \\ v_{n1} & v_{n2} & \cdots & v_{nm} \end{bmatrix} = \begin{bmatrix} w_1 x_{11} & w_2 x_{12} & \cdots & w_m x_{1m} \\ w_1 x_{21} & w_2 x_{22} & \cdots & w_m x_{2m} \\ \vdots & \cdots & \cdots & \vdots \\ w_1 x_{n1} & w_2 x_{n2} & \cdots & w_m x_{nm} \end{bmatrix} \tag{5.8}$$

Equation 5.9 defines the positive ideal solution (PIS) as the maximum value and the negative ideal solution (NIS) as the smallest value for each criterion, while to determine the distance of each alternative to the positive ideal solution (PIS) and negative ideal solution (NIS), Eqs. 5.10 and 5.11 are used.

$$PIS : P^{+} = \left\{ v_1^{+}, v_2^{+}, v_3^{+}, \ldots, v_m^{+} \right\} = \left\{ \left(\max_i v_{ij} | j \in M \right) \right\} \tag{5.9}$$

$$NIS : P^{-} = \left\{ v_1^{-}, v_2^{-}, v_3^{-}, \ldots, v_m^{-} \right\} = \left\{ \left(\min_i v_{ij} | j \in M \right) \right\}$$

$$d_i^+ = \sqrt{\sum_{j=1}^{m} \left(v_{ij} - v_j^+\right)^2}, i = 1, 2, 3, \ldots, n, \text{and } j = 1, 2, 3, \ldots, m. \quad (5.10)$$

$$d_i^- = \sqrt{\sum_{j=1}^{m} \left(v_{ij} - v_j^-\right)^2}, i = 1, 2, 3, \ldots, n, \text{and } j = 1, 2, 3, \ldots, m. \quad (5.11)$$

Ultimately, calculate the coefficient of relative closeness (RC).

$$RC_i = \frac{d_i^-}{d_i^- + d_i^+}, \quad i = 1, 2, 3, \ldots, n \quad (5.12)$$

5.4 Evaluation of the Criteria Employed to Evaluate Anti-ESG Components in Relation to Anticompetition Controversies at European Financial Institutions

In general, the research analysis, which utilized the choice matrix that was generated, can be classified into three primary categories. The subsequent portions provide a detailed explanation of the different periods and their corresponding procedures: initially, the N1 approach is employed to normalize the data in the decision matrix. The objective of this procedure is to standardize the data to a consistent scale, thereby enabling easy comparisons. After Step 1, a matrix that was normalized in accordance with Ni was obtained. Furthermore, the subsequent phase of the analysis entailed the assignment of weights to each criterion *w_ij* and the generation of the weighted normalized matrix in accordance with the procedures delineated in Part 5.3 of the current study. The implementation of the entropy weight TOPSIS technique had a substantial impact on this study. The weights assigned to each criterion for each fiscal year (FY) are presented in Table 5.2.

The implementation of the entropy weight TOPSIS technique revealed that criteria C13, which are explicitly related to environmental issues, have the highest weight and the most significant influence on anticompetition disputes within European financial institutions from 2019 to 2022. Financial institutions play a critical role in the development of a more ecologically sustainable society [115, 167, 194]. They are able to direct markets toward specific objectives by regulating financing and manipulating the value of financial instruments. This encompasses the imposition of limitations on activities that are environmentally detrimental (referred to as "brown" activities) and the encouragement of ecologically sustainable activities (referred to as "green" activities). Banks will assume a significant role in mitigating the risks that arise as bank borrowers encounter increasing challenges from environmental degradation and climate change. European legislators have implemented regulations that mandate that banks incorporate specific objectives and key performance indicators (KPIs) regarding compliance and ethics

Table 5.2 The weight of each criterion for FYs 2022–2018 based on the entropy weight TOPSIS method

Criterion	C1	C2	C3	C4	C5	C6	C7	C8	C9	C10	C11	C12	C13	C14
FY 2022														
ej	– 0.8471	– 0.9498	– 0.9827	– 0.9926	– 0.9985	– 0.9865	– 0.9897	– 0.9985	– 0.9955	– 0.9975	– 0.9995	– 0.9170	– 0.3332	– 0.9661
$d = 1 - ej$	0.1529	0.0502	0.0173	0.0074	0.0015	0.0135	0.0103	0.0015	0.0045	0.0025	0.0005	0.0830	0.6668	0.0339
Wj	0.0267	0.0088	0.0030	0.0013	0.0003	0.0024	0.0018	0.0003	0.0008	0.0004	0.0001	0.0145	0.1165	0.0059
FY 2021														
ej	– 0.8384	– 0.9366	– 0.9771	– 0.9911	– 0.9955	– 0.9892	– 0.9934	– 0.9944	– 0.9955	– 0.9960	– 0.9949	– 0.9095	– 0.4255	– 0.9567
$d = 1 - ej$	0.1616	0.0634	0.0229	0.0089	0.0045	0.0108	0.0066	0.0056	0.0045	0.0040	0.0051	0.0905	0.5745	0.0433
Wj	0.0282	0.0111	0.0040	0.0015	0.0008	0.0019	0.0012	0.0010	0.0008	0.0007	0.0009	0.0158	0.1004	0.0076
FY 2020														
ej	– 0.8164	– 0.9158	– 0.9604	– 0.9762	– 0.9838	– 0.9731	– 0.9811	– 0.9833	– 0.9838	– 0.9844	– 0.9838	– 0.8925	– 0.5823	– 0.9412
$d = 1 - ej$	0.1836	0.0842	0.0396	0.0238	0.0162	0.0269	0.0189	0.0167	0.0162	0.0156	0.0162	0.1075	0.4177	0.0588
Wj	0.0321	0.0147	0.0069	0.0042	0.0028	0.0047	0.0033	0.0029	0.0028	0.0027	0.0028	0.0188	0.0730	0.0103
FY 2019														
ej	– 0.7930	– 0.8913	– 0.9341	– 0.9542	– 0.9629	– 0.9540	– 0.9573	– 0.9623	– 0.9592	– 0.9598	– 0.9623	– 0.8718	– 0.7147	– 0.9154
$d = 1 - ej$	0.2070	0.1087	0.0659	0.0458	0.0371	0.0460	0.0427	0.0377	0.0408	0.0402	0.0377	0.1282	0.2853	0.0846
Wj	0.0362	0.0190	0.0115	0.0080	0.0065	0.0080	0.0075	0.0066	0.0071	0.0070	0.0066	0.0224	0.0499	0.0148
FY 2018														
ej	– 0.7773	– 0.8639	– 0.9227	– 0.9330	– 0.9405	– 0.9311	– 0.9326	– 0.9412	– 0.9369	– 0.9405	– 0.9398	– 0.8519	– 0.7885	– 0.8793
$d = 1 - ej$	0.2227	0.1361	0.0773	0.0670	0.0595	0.0689	0.0674	0.0588	0.0631	0.0595	0.0602	0.1481	0.2115	0.1207
Wj	0.0389	0.0238	0.0135	0.0117	0.0104	0.0120	0.0118	0.0103	0.0110	0.0104	0.0105	0.0259	0.0370	0.0211

(C&E) into their business strategy and risk management. C&E risks are managed and controlled through the establishment of a robust legal framework. Major institutions have implemented measures to illustrate their commitment to sustainability, particularly in relation to climate-related issues, in response to the growing number of legal mandates. The necessity of establishing and implementing compliance and ethics policies, in addition to appropriate key risk indicators (KRIs) and key performance indicators (KPIs), has been emphasized.

However, the measures mentioned above are not implemented to a significant extent by financial institutions. By making considerable investments in the energy and gas sector, financial institutions worldwide, particularly in Europe, contribute significantly to the acceleration of climate change. Financial institutions specialize in offering loans, insurance, and other forms of financing to businesses that operate in the fossil fuel sector. Through this conduct, they support the actions of those who are responsible for the extraction of coal and the combustion of oil, which results in the release of significant quantities of greenhouse gases into the environment. Moreover, financial institutions have the ability to provide financing for specific projects, such as the expansion of a coal mine or the establishment of a new oil platform. These initiatives would be rendered impracticable due to insufficient financial resources. Nevertheless, the consequences of their climate are not restricted to those boundaries. In general, the financial sector provides funding to enterprises that have substantial environmental impacts, including activities such as animal agriculture and weapon manufacturing. British banks have provided financial support to companies that have caused significant environmental damage, such as the extensive deforestation of the Amazon rainforest [195, 196]. The fossil fuel sector is receiving financial assistance from the majority of banks located on the UK's primary retail avenues. Several UK institutions have provided billions of dollars in finance to the fossil fuel sector following the 2015 Paris Agreement, which sought to limit global warming to below 2 degrees Celsius. The Banking on Climate Chaos research specifically identifies certain significant UK high-street banks as being among the top 60 firms that provide financial assistance to the fossil fuel sector. The research encompasses eight financial institutions that are included in the recommendations for ethical current accounts, savings, and mortgages, among other things. Consequently, despite the increasing number of European financial institutions that are emphasizing their dedication to environmental sustainability, these claims frequently fail to withstand a comprehensive examination. Despite the fact that businesses have regulations that explicitly prohibit the use of specific fossil fuels, these regulations frequently contain loopholes that suggest the presence of corruption and anticompetitive behavior.

The European financial industry's analysis revealed that the policy fair competition criteria (C1) had the most significant influence on the escalation of anticompetition disputes in FY 2018. The fair competition policy establishes ethical standards for financial institutions to guarantee equitable competition. Nevertheless, recent research conducted by Chen et al. (2023), Chowdhry (2022), Qi et al. (2022), and Xu et al. (2023) has demonstrated that this standard has resulted in a rise in anticompetition disputes. This suggests that the successive consolidation among financial institutions in Europe has resulted in negative outcomes and a lack of ethical culture. Bank consolidation have occurred in Greece in recent years, particularly during the economic crisis, to safeguard the financial sector's existence. These mergers resulted in a highly concentrated market as a result of the various strategies employed by one business to acquire or exert influence over another. This concentration increased the probability of promoting unethical competition and restricting competition. The consolidation of banks in the Greek financial sector resulted in the establishment of four dominant banks, which considerably reduced the number of banks in the market. The Competition Commission conveyed apprehension regarding these transactions due to their potential to impede competition and reinforce unethical competition practices. K. Wang et al. (2022) and Xia et al. (2022) demonstrated that covert cooperation among administrators of credit institutions' management was a form of unethical competitive techniques.

Conversely, the European banking industry's anticompetition conflicts are minimally affected by the accounting disputes in FY 2018, wage working condition disputes in FY 2019, executive pay disputes in FY 2020–2021, and insider affair disputes in FY 2022. Prior research has demonstrated that regulators augment their supervision and monitoring during periods of crisis, such as the COVID-19 pandemic, in response to significant external disruptions [32, 197–201]. This is particularly relevant to the accounting dispute criteria. Consequently, bank managers may continue to exhibit a strong inclination to manipulate revenue levels during periods of crisis, primarily as a result of their apprehension regarding the severe repercussions of violating laws and their steadfast commitment to achieving financial goals. As a result, they endeavor to refrain from engaging in unconventional or aggressive accounting practices that could potentially lead to discrepancies in financial reporting. The anticompetition conflicts within financial institutions are not influenced by the issue of pay and working conditions. In terms of personnel administration, client assistance, and excellence, these establishments face intense competition [135, 145, 202].

Furthermore, contrary to the current controversy regarding CEO compensation, recent research indicates that it does not have an effect on anticompetition challenges, particularly in the fiscal years 2020 and 2021. Executive pay pertains to instances in which CEOs receive substantial compensation upon their departure from their positions, regardless of whether the organization they oversee is experiencing financial difficulties. The media consistently emphasizes these incidents and portrays them as evidence of a persistent issue [107, 203]. The financial crisis, the COVID-19 pandemic, and the government's intervention to rescue struggling companies using public funds have recently provided a rationale for implementing restrictions on CEO remuneration. It could be contended that government intervention allowed certain CEOs to resign from their positions with substantial quantities of money just prior to the financial collapse of their organizations [204, 205]. Numerous economists have underscored the correlation between the dot-com crisis and the financial and real estate expansions that preceded it over the past 15 years. They contend that the bubble was a direct result of the deliberate actions of central banks, including the implementation of lenient monetary policies and the maintenance of low interest rates. These imprudent monetary policies were most advantageous to those who elected to exercise their stock options during the peak of the market excesses, prior to the subsequent market collapse. Additionally, a moral hazard issue within the financial sector substantially undermined the market's ability to self-correct. This belief was undoubtedly confirmed in most cases as investors were under the impression that the government would provide financial assistance to corporations that were deemed too large to fail [166]. As a result, market discipline deteriorated. This may be primarily attributed to a governmental failure rather than a market failure. The practice of "say on pay" has been implemented by financial organizations to empower shareholders to determine the compensation of the CEO. This method allows shareholders to directly influence compensation decisions through a vote, which may have either a binding or advisory effect, rather than relying solely on the board of directors. Consequently, leaders should refrain from employing strategies that could result in organizational corruption by engaging in unethical competition [80, 206].

Insider trading scandals, which are also referred to as insider deals, may be perceived as a significant indicator of corporate corruption [180, 207, 208]. An entropy weight study illustrates that this component plays a critical role in the facilitation of bribery or deception. The acronym TOPSIS stands for technique for order of preference by similarity to the ideal solution. Anticompetitive practices are minimally affected by this component. The incidence of insider trading or market manipulation may be effectively restricted by the robust and adaptable corporate governance measures of European financial institutions, which may be the cause of this phenomenon. Additionally, research in the disciplines of corporate governance and insider trading has demonstrated that most of the insider trading in publicly listed companies is associated with insufficient internal control and corporate governance. Insider trading is more prevalent in organizations that lack adequate corporate governance. Inadequate corporate governance impedes the ability to establish distinct objectives and norms for business management and provide high-quality information, whereas strong corporate governance may enhance the accuracy and reliability of information for publicly traded companies.

5.5 Evaluation of European Financial Institutions Considering Anti-ESG Factors and Anticompetition Controversies

The TOPSIS methodology was used to assess the performance of European financial institutions in regard to anticompetition problems. This assessment included using the entropy-weight matrix and decision matrix as inputs. The Appendix includes a table that demonstrates the top-performing (+) and bottom-performing (−) subcriteria, as described before. Additionally, we provide the outcomes of the TOPSIS (technology for order preference by similarity to the ideal solution) method, which demonstrate that the most favorable choice exhibits the least deviation from the positive ideal solution and the biggest deviation from the negative ideal solution.

The financial institutions based in Scandinavia that operate in Europe have a greater frequency of anticompetitive disputes, as shown by the data presented in Tables A7 and A8 and the classification determined using the TOPSIS methodology. This finding is rather extraordinary, considering that the Scandinavian banking sector has been under thorough investigation for antitrust infringements for a prolonged duration. Nevertheless, studies in this area have shown the susceptibility of the Nordic nations to become genuine hypocrites when it comes to adherence. However, what was the fundamental cause of this incident? According to the Transparency International Corruption Perceptions Index, Finland, Denmark, and Sweden are ranked as the top nations. Denmark has the top place in the categorization, while Finland and Sweden are ranked in the second and third position as well [209, 210]. Consequently, the inhabitants of these nations are used to the ideals of impartial media, democratic voting, and unrestricted speech. They are not compelled to face corruption often, and the decision-making procedures of their administrations are typically open. Thus, in Scandinavian culture, there is an underlying assumption that people are naturally predisposed to act in an ethical manner.

In the corporate setting, impediments emerge, underscoring the arduousness of the circumstances. The compliance community is unsurprised by the allegations and investigations targeting well-known banks in the Nordic region, including Nordea in Finland, Danske Bank in Denmark, and Swedbank in Sweden, for substantial breaches of money laundering regulations [210]. Our esteemed financial institutions have aided the reintegration of billions of illicit funds into the system. The essential problem is often overlooked in the discourse around the breakdown of compliance procedures in institutions. As per Anagnostopoulos et al. (2022), money laundering is not a leisurely pursuit but rather an essential process for criminals seeking to reintegrate their illegal cash into the financial system [171, 211]. The monies in issue originate from Nordic firms and are used as covert payments to corrupt officials in authoritarian regimes to get government contracts, licenses, and economic successes [171, 212]. Consequently, these variables may lead to a rise in unfair competition among financial institutions in Scandinavia due to the growing number of controversial topics, such as environmental concerns. The latest annual report from the Swedish Climate Policy Council (SCPC), released on March 29, reveals that Sweden is facing a setback in its advancements, while the European Union (EU) is consistently following its current path and even accelerating its development. In addition, Sweden is voicing worry over this scenario. This may be a pivotal element contributing to the heightened environmental conflicts within the financial sector, perhaps leading to the use of anticompetitive tactics to accomplish their objectives.

The findings concerning the nations listed at the bottom of the TOPSIS aggregate score list are quite significant. British financial firms rank poorly in terms of their performance in anticompetition issues. On a theoretical level, the financial service industry in the United Kingdom seems to be fair and just. Competition authorities sometimes do not have a comprehensive understanding of the individual business models of companies, which might differ dramatically from the assumptions found in competition textbooks, especially in financial markets. Moreover, the minimal rates of consumer switching between providers suggest that the retail financial markets often lack demand-side accountability. This undermines the effectiveness of several traditional approaches that aim to promote competition, excluding immoral competition. Financial authorities have determined that the desired amount of competition, which safeguards customers by influencing pricing and quality to match their desires, is not accomplished when trying to address information asymmetry in instances where the demand side is weak.

Chapter 6
A New Paradigm in Long-Term Sustainability: The Quadruple Bottom Line Framework as an Alternative to the Triple Bottom Line

6.1 An Analysis of Sustainability, Historical Trajectories, and Importance

To have a comprehensive understanding of the impact of sustainability and resilience on global society, it is essential to research, study, acknowledge, and conduct a comprehensive review of the subject matter. Ongoing evaluation, modification, and studies are being undertaken to analyze the challenges related to resilience and sustainability. In the context of an organizational structure, "sustainability" refers to a comprehensive examination of social, economic, environmental, and governance issues. At times, sustainability is assessed based on its efficacy, while at other times, it is seen as an additional burden and expense for society. Many local governments worldwide assert that they use sustainability strategies to optimize their operations and mitigate the economic, environmental, and social impacts resulting from their actions. Utilizing both short-term planning and long-term planning to enhance resilience and sustainability may address various challenges that have been identified. The rising energy costs, the political landscape around climate change, and the declining revenue all directly impact the profitability of the present financial situation [213, 214]. Many individuals mistakenly believe that the term "sustainability" denotes a state of long-lasting resilience that is sustained for a significant duration. Moreover, sustainability may be described as the capacity of natural systems to endure, preserve diversity, and remain productive over an extended period.

Both sustainability and resilience prioritize and assign equal significance to environmental, social, and economic factors [88, 215]. Additionally, contemporary enterprises are equally influenced by the widespread acceptance of the idea of sustainability and its interpretation within the context of resilience—both of which are interconnected notions. This attempt is being undertaken to get a more thorough knowledge of the consequences and extent of climate change. A deeper

K. Ragazou et al., *Triple Bottom Line and Multiple Criteria Decision Making Analysis*, Multiple Criteria Decision Making,
https://doi.org/10.1007/978-3-031-78045-5_6

comprehension of sustainability may be attained by using the concept of resilience in situations where the challenges posed by climate change are intensifying. Often, these hurdles manifest as adverse weather events, issues with the infrastructure, and the need for human labor and expertise. Resilience may be defined as an enhanced and refined approach to sustainability planning that specifically addresses the problems posed by climate change. Hitchcock and Willard (2008) state that sustainable organizations are characterized by their proactive efforts to achieve sustainability, their shared understanding of the concept of sustainability, their leadership's support for sustainable practices, and their continuous involvement of key stakeholders [52]. The following are the distinguishing qualities of sustainable organizations in comparison to other forms of organizations. To accomplish this objective, it is essential to prioritize the whole concept and vision of sustainability, including the economy, the environment, and society. More specifically, the establishment of common objectives and specific performance indicators may greatly aid firms in their pursuit of sustained success [216, 217].

6.1.1 The Theoretical Background

To have a thorough grasp of the principles of sustainability and resilience, it is crucial to explore topics closely related to these ideas, given the many viewpoints on sustainability. Gaertner provided a definition for the theory of social choice in 2009, describing it as the study of the collective decision-making process. Furthermore, Gaertner suggested that by amalgamating the preferences of several people, we may get insight into the collective decisions of society. A thorough analysis of sustainability and its associated policies may provide a more complete insight into the activities undertaken and their overall effect on companies or society. According to Elster and Hylland (1989), social choice theory emerged as a result of two distinct issues, one of which involves finding a way to quantify the overall welfare of society [218].

The primary concept of sustainability is to evaluate the effect, relevance, and elusive measurement of social well-being in relation to sustainability. In 1998, Heal presented a thorough elucidation of the underlying concepts of environmental assets. These concepts include considering both current and future elements, acknowledging the constraints imposed by the dynamics of environmental resources, and understanding how environmental resources contribute to economic well-being [219]. This strategy prioritizes environmental advantages as the primary emphasis of sustainability while neglecting the social aspect of the triple bottom line. In 1992, Solow introduced a logical policy approach that used economic theory to justify the idea of potential enhancements to the "economy in relation to its natural resources" [220]. This strategy was deemed to be economically rational. Once again, the focus is on the environment, yet a more favorable environment has a significant impact on both the economy and society. System theory is a theoretical framework that is associated with the examination of sustainability, including the exploration of economic and social choice theories. Von Bertalanffy introduced the notion of a general system theory in 1950. According to him, "general system laws" may be universally applied to every system of a certain kind, regardless of its traits or components.

Every event, irrespective of its association with sustainable energy or any other endeavor, may be seen as an interconnected system with several components. Von Bertalanffy's formulation from 1950 states that generic system theory is relevant to all scientific domains that center on the examination of structure and function. According to Patton (2002), system theory posits that a system is a unified entity that is larger than and distinct from its individual components. This is elucidated within the framework of the theory. This method enables a thorough understanding of the complex sustainability phenomena and the mechanisms that drive the environmental, social, governance, and economic aspects involved. A wide range of ideas, concepts, paradigms, and theories are used to provide a framework for sustainable development and its associated economic, social, and environmental outcomes for enterprises and society. The author of the book "Heal," published in 1998, uses sustainability as a metaphor to illustrate the challenging and significant problems that mankind now faces. A theoretical concept of sustainability may be formulated by drawing from several disciplines, including economics, social science, environmental science, and politics [221–223].

6.1.2 The Starting Point of Sustainability Strategic Planning

Formulating a strategy for sustainability is a complex and intricate task that needs thoughtful deliberation, like the process of strategic planning. The process of developing a strategy for sustainability initiatives involves using an organic, bottom-up, linear, and engaging approach. Companies that have received recognition for their effective sustainability and resilience strategies display commitment by incorporating sustainability into their aims, concepts, and values. They do this via strategic planning and by allocating funding accordingly. This is done to showcase their unwavering commitment to the principle of sustainability [39, 224, 225]. Companies have the capability to convert their planning into a strategy that is both achievable and capable of increasing value and then implement it successfully, leading to advantages for the whole community. This is achieved by implementing sustainability and resilience planning. The integration of sustainability into all facets of municipal governance is a time-consuming endeavor that cannot be completed in isolation. The successful implementation of sustainability initiatives is influenced by a variety of variables, such as, but not limited to, the following:

- Lean concepts are implemented to optimize operational performance and improve internal efficiency.
- The sustainability plan, renewable energy objectives, and green building strategy are among the most critical phases of policy and planning.

- The community's active engagement and partnership are important in the pursuit of sustainability.
- It is imperative that key stakeholders and personnel, regardless of their status as elected or appointed officials in the municipal council, participate actively.
- Provide employees with the ability to advocate for and maintain sustainable objectives.
- Utilize sustainability progress reports to evaluate, supervise, and disseminate the outcomes of sustainability-related initiatives. It is imperative to establish precise targets for the strategy as they are critical components of these duties, in addition to identifying clearly defined goals and objectives.
- Coordinate objectives with budgetary objectives and fiscal policies. It is imperative to incorporate these into such endeavors, as well as establish explicit objectives and aims.
- Connect objectives to budgetary objectives and fiscal policies.

6.1.3 Resilience Planning

The issue of climate change is very relevant and can be a major obstacle for companies. The failure of national governments to respond has resulted in certain municipal governments adopting independent measures to solve this pressing and potentially life-threatening situation. However, there is a consensus among scientists about the origins and implications of climate change. A study conducted in 2014 by the Intergovernmental Panel on Climate Change (IPCC) and the United States Global Change Research Program (USGCRP) indicates that companies worldwide, including those in the United States, are exposed to considerable risks due to climate change. The Kyoto Protocol, agreed upon by the United Nations Framework Convention on Climate Change (UNFCCC) in 1997, is a global agreement that mandates member governments to restrict their emissions of greenhouse gases [226, 227]. The United Kingdom opted to abstain from participation in the deal, while more recently, the United States government withdrew from the already ratified Paris climate treaty. The ruling has led to the shifting of the responsibility of addressing the repercussions of inaction to the communities and governments of the United States.

The US Department of State formally announced its decision to withdraw from the agreement on August 4, 2017. Nevertheless, the United States has affirmed its commitment to actively engage in current and future international discussions and conventions about climate change. The United States' inadequate leadership in tackling the difficulties and consequences of climate change is harmful to several communities struggling with this issue. The primary catalyst for climate change is the accumulation of greenhouse gases in the atmosphere, which has been steadily rising in recent years due to an increased dependence on fossil fuels. Based on the USCCSP's (2009) conclusions, climate scientists generally agree that the Earth's climate is experiencing substantial alterations, mostly due to human activities that have resulted in increased amounts of greenhouse gas emissions. The primary contributors to these emissions, including carbon dioxide, methane, and nitrous oxide, are the combustion of fossil fuels and alterations in land use. Measurements have shown a 25% rise in these emissions throughout the last century. The evident manifestation of global warming in the climate system presents significant challenges in effectively forecasting the extensive ramifications of climate change in various regions, mostly owing to the restricted availability of data in terms of both time and place. The empirical data unequivocally demonstrates that human activities are the predominant catalysts for the fast escalation of greenhouse gas emissions.

In 2016, the National Aeronautics and Space Administration (NASA) issued a definitive and official warning regarding the consequences of climate change. These include rising temperatures, extended frost-free periods, alterations in precipitation patterns, heightened occurrences of droughts and heatwaves, intensified storms, a projected sea level increase of 1–4 feet by 2100, and a significant likelihood of the Arctic being ice-free. The main goal is to focus on localizing the impacts of climate change by offering detailed recommendations to improve the community's capacity to endure and adjust to these outcomes. Furthermore, the objective is to support firms in their efforts to adapt to climate change and enhance their ability to recover from disasters and build resilience [14, 228]. Climate resilience encompasses the ability to withstand, recover from, and adjust to the adverse effects of climate change, such as changes in the environment and heightened disruptions to the economy and society. Businesses need crucial preventive tools and resources to swiftly address and mitigate disasters and shocks. To establish a resilient society, it is imperative that all essential systems have the capacity to adapt, tolerate alterations, and maintain vigilance. Resilience refers to the capacity of both human populations and the natural environment to withstand and bounce back from potential hardships, enabling communities to effectively address future challenges [228].

In its 2013 report, the West Michigan Environmental Action Council offered a thorough and comprehensive examination of the significant and far-reaching consequences of climate change. Given the connections that exist across sectors and the overall operation of the system, climate change has a significant impact on every sector. In order to cultivate resilience, it is imperative to possess a thorough comprehension of the community's needs, the substantial connections that exist among the various sectors, and the ability of these sectors to meet those needs in a constantly changing environment. In addition to their emphasis on sustainability, businesses are employing resilience planning to specifically address the challenges associated with climate change and the recurrence of severe weather events. It involves the resolution of issues related to the economy, society, the environment, and social governance. Adaptation to climate change and mitigation of climate change are both aspects of these programs. To achieve their objectives, businesses should integrate all aspects of their operations into comprehensive sustainability and resilience strategies. It is imperative that companies include the identification of the risks posed by climate change in their risk assessment and planning processes to promote the development of sustainability and resilient communities [229, 230].

Businesses implement both strategies due to the importance of climate adaptation and mitigation initiatives. When designing for resilience, it is crucial to consider all aspects of the built environment. In the same way that strategies that attempt to mitigate the impact of a problem may be perceived as methods to adapt to the situation, the converse is also accurate. The evaluation of emergency plans to accommodate individuals who are susceptible to climate-related hazards, the development of environmentally friendly infrastructure in addition to the improvement and expansion of existing conventional infrastructure, and the preparedness to address unforeseen disasters should be included in the planning process. These methods should be integrated into the operations of business administrators to reduce the negative effects of heat islands, improve air quality, reduce water usage, and reduce energy consumption.

6.2 Sustainability in the Private Sector

Sustainability is not just limited to the public sector. According to Werbach (2009), there are major and fast changes happening in the global economy, ecology, and political institutions [59]. These changes need an explanation for why corporations are adopting sustainable practices. In 2009, Waddock performed an extensive study on "emerging corporate practices" that benefit organizations in reaching their sustainability objectives [59]. These tasks include the fair distribution of carbon costs, accountability for products, and the management of the whole life cycle. They also include comparing performance with other firms. Comparing performance versus other companies is an extra approach. Sustainability refers to the implementation of long-term strategies and actions by a corporation to supply services and commodities in a way that is both ecologically friendly and efficient. Sustainable organizations aim to maximize their beneficial economic and social influence while simultaneously reducing the adverse environmental effects of their activities. Businesses are aware of the significance of sustainability and resilience in comprehending the possible risks linked to declining infrastructure, communities, cities, and society.

Aligašić (2018b) states that the individuals involved possess an understanding of the potential adverse effects that these threats might have on their actions and, therefore, their financial outcomes. By integrating principles of sustainability, companies may quickly adapt and explore potential enduring sustainable solutions. There seems to be a connection between the challenges faced by large-scale communities and businesses and the advantages and disadvantages associated with the global economy. Enterprises, municipal governments, and other organizations use sustainability and resilience as strategic methods to achieve their long-term objectives. This is happening due to the ongoing dissemination of cultural norms. The purpose of this framework is to provide corporations with a diverse array of models, approaches, and theories to aid them in addressing complex social and environmental sustainability issues. Hardjono et al. (2004) proposed a comprehensive approach to incorporate sustainability into enterprises through the use of organizational dynamics. The authors used the symbolic interaction technique in a methodical manner to facilitate the adoption of corporate sustainability and social responsibility programs inside enterprises. In 2004, Brownsword and his colleagues performed a study in which they devised a reliable method to evaluate the availability and need for environmentally friendly energy in metropolitan areas. This paradigm considers It does this by considering the technical and socioeconomic barriers related to energy use in residential and commercial settings [224, 231].

The results of this inquiry are included into the process to develop a model for energy management in urban environments. A distinctive feature of this study is this component's contribution to assessing sustainable energy and its alignment with sustainable cities. It is also one of the novel components of this research. This chapter offers a comprehensive exploration of the practical consequences that firms must consider in relation to energy and sustainability planning. Additionally, it presents an assessment of the influence that sustainability has on organizations. Some individuals hold the viewpoint that organizations, especially corporations, use the notion of sustainability to advance social causes, such as environmental preservation and climate change mitigation, and to deviate from conventional business practices. In their study, Bennell and Kearnis (2005) investigated how firms used their economic and corporate influence to promote sustainability issues, such as climate change, and to accomplish political goals. Furthermore, these challenges have broader ramifications for society, in addition to their impact on the markets. To attain sustainability, a coordinated and cooperative endeavor is needed, which involves the commitment of a wide range of persons, including nonprofit, governmental, and corporate groups. Savitz and Weber proposed the notion of a triple bottom line in their 2006 book [232]. According to them, the triple bottom line is the primary determinant of a company's sustainability. The authors argue that a sustainable company may successfully handle its risks and take advantage of its possibilities by recognizing important nonfinancial stakeholders and engaging them in areas of shared interest [232]. This is a contention that is put up in the authors' piece of writing. Businesses have the obligation to assess the social and environmental consequences of their activities, not just for economic reasons but also because it may be mutually advantageous and could be a lucrative business prospect [233]. To meet these criteria, it is necessary for firms committed to sustainability to successfully complete this examination [222, 223].

6.3 Sustainable Development and the Triple Bottom Line

The initial organization to establish the foundation for sustainable development and sustainability was the United Nations' World Commission on Environment and Development. In 1987, the expression "sustainable development" was introduced by the United Nations' World Commission on Environment and Development. In addition to guaranteeing that the needs of future generations are satisfied, enterprises and individuals should ensure the responsible and efficient utilization of resources. The Brundtland Commission, which is more commonly known as the World Commission on Environment and Development, introduced the concept of sustainability in its report "Our Common Future." "Sustainable development" is the process of satisfying the requirements of the present generation without compromising the capacity of future generations to meet their own needs. The two main concepts are as follows: a transformative process that ensures that decisions on investments, technological advancements, institutional modifications, and resource use are in accordance with the human requirements and aspirations of the current and future generations.

While in its early phases of defining sustainability, the United Nations' World Commission on Environment and Development established the groundwork for sustainable development in 1987. The framework's primary focus was on future societal needs. Nevertheless, a substantial quantity of research was conducted to enhance and fortify the concept of sustainability following the initial development of the sustainability platform. Diverse perspectives regarding the potential repercussions of sustainability may exist among stakeholders, including enterprises, communities, and society. Stubbs and Cocklin (2008) argue that there is a lack of agreement on the exact definition of sustainability and that there is a disagreement over the idea [234]. Elkington's book in 1997 presented groundbreaking and pioneering viewpoints on sustainability to the global audience. He urged corporations to evaluate the ramifications of their activities on society and the environment [233]. Elkington's 1997 article titled "Cannibals with Forks: The Triple Bottom Line of 21st Century Business" played a crucial role in introducing the notion of the triple bottom line and its potential influence on enterprises and organizations. The triple bottom line (TBL) is a commonly used phrase to describe the three areas of impact, as defined by Elkington (1997) and Savitz and Webber (2006). It is a concept within the framework of sustainability that has the goal of maintaining environmental ecological integrity, promoting social fairness, and achieving economic success [2, 233].

6.4 A Shift from TBL to the QBL Framework for Sustainability

During the initial phases of sustainability development, the triple bottom line method was implemented. The objective of this methodology was to evaluate the social influence of organizations. It is a fair approach that elucidates the extent to which a company provides value to both its shareholders and society, as Savitz and Weber (2006) have noted [232]. The triple bottom line is a strategy that is equitable. It was developed as a framework for evaluating and quantifying the contributions that companies make to society and the environment, in addition to the economic benefits they provide, as per Elkington (1997). The capacity of organizations to advance sustainability is acknowledged in the context of societal progress. Elkington (2012) further argued in his subsequent works that sustainability serves as a foundation for improved corporate governance, resulting in the establishment of "genuinely sustainable capitalism." To achieve success, private sector corporations must establish objectives that prioritize economic development, environmental conservation, and social equity, as specified by the triple bottom line. While Elkington's (1997) triple bottom line concept is primarily focused on the private sector, it can also be applied to public sector organizations to achieve comparable benefits through its comprehensive principles [233]. Ongoing innovation and review are necessary as the concept of sustainability, which is founded on three fundamental pillars, continues to develop. The quadruple bottom line paradigm, which is proposed, examines the concept of sustainability through a governance perspective.

The quadruple bottom line's expanded version represents an organization's capacity to incorporate and execute a comprehensive array of policies and initiatives that address the economic, social, environmental, and governance aspects of sustainability. Community engagement to guarantee effective service delivery, fiscal restraint and resilience, and the principles of accountability and transparency are the distinguishing characteristics of governance. It is imperative to adopt a dynamic governance model in order to achieve resilience and sustainability. Governance that is effective is essential for the advancement of resilience and sustainability.

6.5 Stages of Sustainability Planning and the Quadruple Bottom Line (QBL)

It is essential to guarantee that sustainability planning is consistent with the fiscal year, the budget process, and the establishment of benchmarks and baseline years. Furthermore, the extent to which organizations can endure and surmount challenges over an extended period is contingent upon the alignment of their benchmark data and objectives with their current strategies and policies [4, 235]. During a period in which organizations, communities, and climate-related projects are prevalent, the plan should explicitly define the altering sustainability objectives, fundamental values, primary objectives, strategies, and transformative initiatives. To guarantee that local government officials are capable of effectively confronting forthcoming challenges and opportunities in a synchronized and pragmatic manner, the sustainability plan should explicitly specify the objectives and goals. The sustainability strategy is designed to enhance the triple bottom line framework by integrating a quadruple bottom line (QBL) framework. Additionally, it enhances the strategies of other municipalities.

The quadruple bottom line (QBL) is the foundation for this methodology, which encompasses economic, social, environmental, and governance components. The precise objectives of the plan can be broken down into four overarching themes, with the specific targets further divided into ten distinct goals: (1) economic opportunity, (2) high-quality neighborhood, (3) social fairness, (4) secure community, (5) resilient systems, (6) well-balanced transportation, (7) sustainable resources, (8) financial resilience, (9) transparency and accessibility, and (10) effective governance. The new sustainability plan is a valuable addition to the extant plans of cities that have already implemented one. The triple bottom line (TBL) framework is the foundation for the preponderance of the initiatives. However, the sustainability strategy can be effectively addressed by incorporating a quadruple bottom line (QBL) structure, which can be used to facilitate governance and related initiatives. Additionally, the city guarantees the uninterrupted and consistent functioning of its operations by sustaining a high level of honesty and integrating current plans in accordance with its objectives, outcomes, and themes. The quadruple bottom line is the collective term used to describe the economic, social, environmental, and governance components of this strategy. By adhering to one of the four fundamental principles of QBL, prospective subjects may develop a comprehensive sustainability strategy for a community. The classification of each specific purpose into multiple objectives and outcomes is closely associated with the city's financial and fiscal policies.

It is imperative to prioritize accountability, transparency, and accessibility in order to ensure the sustainability and resilience of local administrations and communities. The decisions, behaviors, and operations of organizations, citizens, institutions, and enterprises significantly influence the capacity of a community to sustain and preserve its long-term viability. Designated officials and employees are accountable for actively engaging in the pursuit of sustainability objectives. In addition, the appointed champions are charged with the responsibility of overseeing and carrying out the sustainability goals that are defined in this plan. This strategy results in an increase in the efficacy of each target by fostering a sense of personal responsibility and ownership. Companies have implemented a sustainability and resilience plan to enhance operational efficiency and ensure good governance. To address the issue of affordable housing opportunities, promote diversity and inclusivity, decrease energy consumption, increase the generation of renewable energy, minimize waste, facilitate community outreach initiatives, standardize system procedures, and collaborate with organizations, companies, and citizens in order to promote growth while preserving natural resources, the plans can be implemented. The authorities of the municipality are responsible for overseeing the progress of each target and producing a document that is available to the public and that details the strategies and activities that have been executed to attain the quadruple bottom line (QBL). Reports on the progress made in terms of sustainability and resilience are made available to the public on an annual or biannual basis.

6.6 Assessing the Quadruple Bottom Line Metrics

The triple bottom line (TBL) is a concept that is often used by businesses in order to evaluate efforts that are relevant to sustainability and resilience. There are economic, social, and environmental considerations that are included in the TBL. In general, it is not possible to classify each target into a single category since most targets belong to a number of different categories. It is common for local governments to fail to monitor and keep track of the fundamental aspects of sustainable planning, which include transparency, accountability, community participation, and good governance. As a result of the relevance of these components in the planning process, this presents a difficult conundrum. During his presentation in 2017, Alibašić provided a comprehensive and extensive description of the process of sustainability planning, with a particular emphasis on the quadruple bottom line (QBL). The author placed a strong emphasis on the need for businesses to put in place certain policies and programs to meet the economic, social, environmental, and governance components of sustainability. As stated by Aligašić (2017), governance is a notion that encompasses a multitude of traits, including but not limited to fiscal responsibility, participation, community involvement, transparency, answerability, and accountability. A thorough investigation into the many indications that make up the quadruple bottom line is necessary [236, 237].

References

1. da Silva de Camargo Barros, R., & Florêncio de Almeida, L. (2024). Triple-A approach and global value chain governance (GVC): The case of Brazilian beef. *Journal of Rural Studies, 107*, 103241. https://doi.org/10.1016/J.JRURSTUD.2024.103241
2. Agnihotri, A., Bhattacharya, S., & Gorton, M. (2024). Founder's birth order and triple bottom line in B2B SMEs. *Industrial Marketing Management, 117*, 1–13. https://doi.org/10.1016/J.INDMARMAN.2023.12.010
3. Huang, X., Ullah, M., Wang, L., et al. (2024). Green supply chain management practices and triple bottom line performance: Insights from an emerging economy with a mediating and moderating model. *Journal of Environmental Management, 357*, 120575. https://doi.org/10.1016/j.jenvman.2024.120575
4. Schaltegger, S., Lüdeke-Freund, F., & Hansen, E. G. (2012). Business cases for sustainability: The role of business model innovation for corporate sustainability. *International Journal of Innovation and Sustainable Development, 6*, 95–119.
5. Wu, B., Fang, H., Jacoby, G., et al. (2022). Environmental regulations and innovation for sustainability? Moderating effect of political connections. *Emerging Markets Review, 50*, 100835. https://doi.org/10.1016/j.ememar.2021.100835
6. Sohns, T. M., Aysolmaz, B., Figge, L., & Joshi, A. (2023). Green business process management for business sustainability: A case study of manufacturing small and medium-sized enterprises (SMEs) from Germany. *Journal of Cleaner Production, 401*, 136667. https://doi.org/10.1016/J.JCLEPRO.2023.136667
7. Elliot, S. (2011). Transdisciplinary perspectives on environmental sustainability: A resource base and framework for IT-enabled business transformation. *MIS Quarterly, 35*, 197–236. https://doi.org/10.2307/23043495
8. Benkert, J. (2021). Reframing business sustainability decision-making with value-focussed thinking. *Journal of Business Ethics, 174*, 441–456. https://doi.org/10.1007/S10551-020-04611-4
9. Martínez-Bravo, M. d. M., Capobianco-Uriarte, M. d. l. M., Terán-Yépez, E., et al. (2024). Integrating sustainability into business and management studies in higher education. *International Journal of Management Education, 22*, 100939. https://doi.org/10.1016/j.ijme.2024.100939
10. Murcia, M. J., & Acosta, P. (2023). Accounting for plural cognitive framings of growth and sustainability: Rethinking management education in Latin America. *Journal of Business Ethics, 185*, 299–313. https://doi.org/10.1007/S10551-022-05180-4

K. Ragazou et al., *Triple Bottom Line and Multiple Criteria Decision Making Analysis*, Multiple Criteria Decision Making,
https://doi.org/10.1007/978-3-031-78045-5

11. Cheng, J., Mahinder Singh, H. S., Zhang, Y. C., & Wang, S. Y. (2023). The impact of business intelligence, big data analytics capability, and green knowledge management on sustainability performance. *Journal of Cleaner Production, 429*, 139410. https://doi.org/10.1016/j.jclepro.2023.139410
12. Utama, D. R., Hamsal, M., Rahim, R. K., & Furinto, A. (2024). The effect of digital adoption and service quality on business sustainability through strategic alliances at port terminals in Indonesia. *Asian Journal of Shipping and Logistics, 40*, 11–21. https://doi.org/10.1016/j.ajsl.2023.12.001
13. Gomes, S., & Pinho, M. (2023). Can we count on the commitment of European SMEs to achieve SGD12? An exploratory study of business sustainability. *Journal of Cleaner Production, 425*, 139016. https://doi.org/10.1016/J.JCLEPRO.2023.139016
14. Wang, A., Rauf, A., Ozturk, I., et al. (2024). The key to sustainability: In-depth investigation of environmental quality in G20 countries through the lens of renewable energy, economic complexity and geopolitical risk resilience. *Journal of Environmental Management, 352*, 120045. https://doi.org/10.1016/j.jenvman.2024.120045
15. Ragazou, K., Passas, I., & Sklavos, G. (2022). Investigating the strategic role of digital transformation path of SMEs in the era of COVID-19: A bibliometric analysis using R. *Sustainability (Switzerland), 14*, 11295. https://doi.org/10.3390/SU141811295
16. Kobrin, S. J. (1976). Foreign direct investment, industrialization, and social change. *Journal of Conflict Resolution, 20*, 497–522. https://doi.org/10.1177/002200277602000306
17. Zhou, B., Gao, K., Ning, J., et al. (2024). Does air pollution change the business strategy for investment? The strategic influence of national governance. *Finance Research Letters, 63*, 105302. https://doi.org/10.1016/J.FRL.2024.105302
18. Iqbal, N., Umar, Z., Ruman, A. M., & Jiang, S. (2024). The term structure of yield curve and connectedness among ESG investments. *Research in International Business and Finance, 67*, 102145. https://doi.org/10.1016/j.ribaf.2023.102145
19. Ma, Y., & Hu, X. (2024). Shadow banking and SME investment: Evidence from China's new asset management regulations. *International Review of Economics and Finance, 93*, 332–349. https://doi.org/10.1016/J.IREF.2024.03.033
20. Souillard, B. (2022). Intra-industry spillovers of profit shifting and investments in tax havens. *Journal of Economic Behavior and Organization, 204*, 581–599. https://doi.org/10.1016/j.jebo.2022.09.021
21. Jadhav, P., & Katti, V. (2012). Institutional and political determinants of foreign direct investment: Evidence from BRICS economies. *Poverty & Public Policy, 4*, 49–57. https://doi.org/10.1002/POP4.5
22. Johnston, W. J., & Cortez, R. M. (2023). Business-to-business digitalization, artificial intelligence, and social action. *Journal of Business Research, 172*, 113952. https://doi.org/10.1016/j.jbusres.2023.113952
23. Haski-Leventhal, D., Pournader, M., & Leigh, J. S. A. (2022). Responsible management education as socialization: Business students' values, attitudes and intentions. *Journal of Business Ethics, 176*, 17–35. https://doi.org/10.1007/S10551-020-04593-3
24. Fitriani, N., Setiawan, D., Aryani, Y. A., & Arifin, T. (2023). Does social media affect performance in e-commerce business? The role of customer management. *Journal of Open Innovation: Technology, Market, and Complexity, 9*, 100171. https://doi.org/10.1016/j.joitmc.2023.100171
25. Habib, A. M. (2023). Do business strategies and environmental, social, and governance (ESG) performance mitigate the likelihood of financial distress? A multiple mediation model. *Heliyon, 9*, e17847. https://doi.org/10.1016/j.heliyon.2023.e17847
26. Ardito, L., Capolupo, P., De Massis, A., et al. (2024). Unveiling environmental, social, and governance dynamics in family firms. *Journal of Family Business Strategy*, 100608. https://doi.org/10.1016/J.JFBS.2024.100608
27. Malinauskaite, J., & Jouhara, H. (2024). Corporate social responsibility (CSR) and environmental, social and governance (ESG). Introduction to sustainable business models.

Sustainable Energy Technology, Business Models, and Policies, 41–66. https://doi.org/10.1016/B978-0-443-18454-3.00010-2
28. Malinauskaite, J., & Jouhara, H. (2024). Corporate social responsibility (CSR) and environmental, social and governance (ESG). Introduction to sustainable business models. *Sustainable Energy Technology, Business Models, and Policies*, 41–66. https://doi.org/10.1016/B978-0-443-18454-3.00010-2
29. Wu, B., Wang, Q., Fang, C. H., et al. (2022). Capital flight for family? Exploring the moderating effects of social connections on capital outflow of family business. *Journal of International Financial Markets, Institutions and Money, 77*, 101491. https://doi.org/10.1016/j.intfin.2021.101491
30. Cui, X., Wang, X., & Feng, Y. (2019). Examining urban metabolism: A material flow perspective on cities and their sustainability. *Journal of Cleaner Production, 214*, 767–781. https://doi.org/10.1016/j.jclepro.2019.01.021
31. Apostol, O., Mäkelä, M., Heikkilä, K., et al. (2021). Triggering sustainability communication in a B2B context: Combining action research and sensemaking. *Accounting, Auditing and Accountability Journal, 34*, 849–876. https://doi.org/10.1108/AAAJ-08-2019-4125/FULL/HTML
32. Zhou, Y., Li, X., Wang, X., & Yuen, K. F. (2022). Intelligent container shipping sustainability disclosure via stakeholder sentiment views on social media. *Marine Policy, 135*, 104853. https://doi.org/10.1016/j.marpol.2021.104853
33. Yao, N., Fabus, M., Hu, L., & Qian, F. (2023). Resource efficiency and economic sustainability in APEC: Assessing the financial sector's role. *Resources Policy, 85*, 103963. https://doi.org/10.1016/j.resourpol.2023.103963
34. Salim, K., Disli, M., Ng, A., et al. (2023). The impact of sustainable banking practices on bank stability. *Renewable and Sustainable Energy Reviews, 178*, 113249. https://doi.org/10.1016/J.RSER.2023.113249
35. Jum'a, L., Zimon, D., & Ikram, M. (2021). A relationship between supply chain practices, environmental sustainability and financial performance: Evidence from manufacturing companies in Jordan. *Sustainability, 13*, 2152. https://doi.org/10.3390/SU13042152
36. Aras, G., Tezcan, N., Kutlu Furtuna, O., & Hacioglu Kazak, E. (2017). Corporate sustainability measurement based on entropy weight and TOPSIS: A Turkish banking case study. *Meditari Accountancy Research, 25*(3), 391–413. emerald.com
37. El Bilali, H., Strassner, C., & Ben Hassen, T. (2021). Sustainable agri-food systems: Environment, economy, society, and policy. *Sustainability, 13*, 6260. https://doi.org/10.3390/SU13116260
38. Laasch, O., Moosmayer, D. C., & Antonacopoulou, E. P. (2023). The interdisciplinary responsible management competence framework: An integrative review of ethics, responsibility, and sustainability competences. *Journal of Business Ethics, 187*, 733–757. https://doi.org/10.1007/S10551-022-05261-4
39. Ramos, L., Bermeo, M., Quelal, A., et al. (2024). SustainMeter: A knowledge-based system for assessing sustainability of SMEs in the fashion industry. *SoftwareX, 25*, 101611. https://doi.org/10.1016/J.SOFTX.2023.101611
40. Appiah-Kubi, E. (2024). Management knowledge and sustainability reporting in SMEs: The role of perceived benefit and stakeholder pressure. *Journal of Cleaner Production, 434*, 140067. https://doi.org/10.1016/J.JCLEPRO.2023.140067
41. Harris, R. (2010). Evaluating Internet research sources. *Virtual Salt, 17*(1), 1–17.
42. Keeney RL, Raiffa H (1993) Decisions with multiple objectives: Preferences and value trade-offs. *Decisions with multiple objectives*. https://doi.org/10.1017/CBO9781139174084.
43. Hwang, C.-L., & Yoon, K. (1981). *Methods for multiple attribute decision making* (pp. 58–191). https://doi.org/10.1007/978-3-642-48318-9_3
44. Mohamed Mouhoumed, R., Ekmekcioğlu, Ö., & Özger, M. (2023). A hybrid MCDA approach for delineating sites suitable for artificial groundwater recharge using drywells. *Journal of Hydrology, 620*, 129387. https://doi.org/10.1016/j.jhydrol.2023.129387

45. Withanage, W. K. N. C., Gunathilaka, M. D. K. L., Mishra, P. K., et al. (2023). Indexing habitat suitability and human-elephant conflicts using GIS-MCDA in a human-dominated landscape. *Geography and Sustainability, 4*, 343–355. https://doi.org/10.1016/j.geosus.2023.08.004
46. Mvula, R. L. S., Mundike, J., & Nguvulu, A. (2023). Spatial suitability analysis for site selection of municipal solid waste landfill using hybrid GIS and MCDA approach: The case of Kitwe, Zambia. *Scientific African, 21*, e01885. https://doi.org/10.1016/j.sciaf.2023.e01885
47. DasGupta, R., & Roy, A. (2023). Firm environmental, social, governance and financial performance relationship contradictions: Insights from institutional environment mediation. *Technological Forecasting and Social Change, 189*, 122341. https://doi.org/10.1016/j.techfore.2023.122341
48. Salvi, A., Nirino, N., Enrico, B., & Igor, G. (2024). Payout policy and ESG: A European investigation. *Research in International Business and Finance, 68*, 102189. https://doi.org/10.1016/j.ribaf.2023.102189
49. Loucks, D. P., & van Beek, E. (2017). *Water resource systems planning and management: An introduction to methods, models, and applications* (pp. 1–624). Springer. https://doi.org/10.1007/978-3-319-44234-1/COVER
50. Keeney, R. L., & Raiffa, H. (1979). Decisions with multiple objectives: Preferences and value trade-offs. *IEEE Transactions on Systems, Man, and Cybernetics, 9*, 403. https://doi.org/10.1109/TSMC.1979.4310245
51. Aberilla, J. M. (2024). Life cycle sustainability assessment: Integrating life cycle analysis and decision analysis. In *Encyclopedia of sustainable technologies* (pp. 274–280). https://doi.org/10.1016/B978-0-323-90386-8.00072-3
52. Hitchcock, D. E., Willard, M. L., & AtKisson, A. (2008). *The step-by-step guide to sustainability planning* (1st ed.). Earthscan.
53. Maznan, N. A., Jaapar, A., Bari, N. A. A., & Zawawi, M. (2012). Value management: Private sector's perception. *Procedia - Social and Behavioral Sciences, 50*, 383–391. https://doi.org/10.1016/j.sbspro.2012.08.043
54. Lin, X., Mazlan, A. N., Ismail, S., et al. (2022). Status of value management studies in construction projects: A systematic review. *Ain Shams Engineering Journal, 14*, 101820. https://doi.org/10.1016/j.asej.2022.101820
55. Zhang, Z., Liu, J., Xiao, C., & Chen, G. (2023). Making waves: Enhancing sustainability and resilience in coastal cities through the incorporation of seawater into urban metabolism. *Water Research, 242*, 120140. https://doi.org/10.1016/j.watres.2023.120140
56. Valencia, A., Zhang, W., & Bin, C. N. (2022). Sustainability transitions of urban food-energy-water-waste infrastructure: A living laboratory approach for circular economy. *Resources, Conservation and Recycling, 177*, 105991. https://doi.org/10.1016/j.resconrec.2021.105991
57. Minton, E. A., Tan, S. J., Tambyah, S. K., & Liu, R. L. (2022). Drivers of sustainability and consumer well-being: An ethically-based examination of religious and cultural values. *Journal of Business Ethics, 175*, 167–190. https://doi.org/10.1007/S10551-020-04674-3
58. Han, Z., Xie, W., Yu, H., et al. (2024). Rethinking industrial land-use in American rust cities towards sustainability based on a block-level model. *Journal of Environmental Management, 352*, 120067. https://doi.org/10.1016/j.jenvman.2024.120067
59. Finisterra Do Paço, A. M. (2009). Adam Werbach, strategy for sustainability. A business manifesto. *International Review on Public and Nonprofit Marketing, 6*, 187–188. https://doi.org/10.1007/S12208-009-0039-4
60. Hu, N., & Ahmad, U. S. (2024). The impact of green credit legislation on business financing: Insights from Chinese polluting firms. *Heliyon, 10*, e32722. https://doi.org/10.1016/J.HELIYON.2024.E32722
61. Qi, R., Liu, C., & Zhang, Q. (2024). Evidence mechanism for promoting natural resources and green recovery through green economy. *Resources Policy, 91*, 104760. https://doi.org/10.1016/j.resourpol.2024.104760

62. Xu, J., & Xu, W. (2024). Financing sustainable smart city projects: Public-private partnerships and green bonds. *Sustainable Energy Technologies and Assessments, 64*, 103699. https://doi.org/10.1016/j.seta.2024.103699
63. Wang, W., Huang, H., Peng, X., et al. (2024). Utilizing support vector machines to foster sustainable development and innovation in the clean energy sector via green finance. *Journal of Environmental Management, 360*, 121225. https://doi.org/10.1016/j.jenvman.2024.121225
64. Bianchi, R. V., McCabe, S., Cisneros-Martínez, J. D., & Andrés Domínguez-Gómez, J. (2023). Contested discourses in social tourism: A relational political economy perspective. *Annals of Tourism Research, 102*, 103640. https://doi.org/10.1016/j.annals.2023.103640
65. Kim, S., Yoon, Y. U., & Kim, B. C. (2024). Method for estimating management points for dimensional quality management in shipbuilding and offshore plant construction using terrestrial laser scanning and a computer-aided design model. *Advanced Engineering Informatics, 62*, 102591. https://doi.org/10.1016/j.aei.2024.102591
66. Ma, R., Mao, D., Cao, D., et al. (2024). From vineyard to table: Uncovering wine quality for sales management through machine learning. *Journal of Business Research, 176*, 114576. https://doi.org/10.1016/j.jbusres.2024.114576
67. Scottish Funding Council. (2020). *Coherence and Sustainability: A review of Scotland's Colleges and Universities.*
68. Feeney, M., Grohnert, T., Gijselaers, W., & Martens, P. (2023). Organizations, learning, and sustainability: A cross-disciplinary review and research agenda. *Journal of Business Ethics, 184*, 217–235. https://doi.org/10.1007/S10551-022-05072-7
69. Hansen, E. G., & Schaltegger, S. (2016). The sustainability balanced scorecard: A systematic review of architectures. *Journal of Business Ethics, 133*, 193–221. https://doi.org/10.1007/s10551-014-2340-3
70. D'Apolito, E., Galletta, S., Iannuzzi, A. P., & Labini, S. S. (2024). Sustainability and bank credit access: New evidence from Italian SMEs. *Research in International Business and Finance, 69*, 102242. https://doi.org/10.1016/J.RIBAF.2024.102242
71. Kelling, N. K., Sauer, P. C., Gold, S., & Seuring, S. (2021). The role of institutional uncertainty for social sustainability of companies and supply chains. *Journal of Business Ethics, 173*, 813–833. https://doi.org/10.1007/S10551-020-04423-6
72. Bilyay-Erdogan, S., Danisman, G. O., & Demir, E. (2023). ESG performance and dividend payout: A channel analysis. *Finance Research Letters, 55*, 103827. https://doi.org/10.1016/j.frl.2023.103827
73. Wang, H., Jiao, S., Bu, K., et al. (2023). Digital transformation and manufacturing companies' ESG responsibility performance. *Finance Research Letters, 58*, 104370. https://doi.org/10.1016/j.frl.2023.104370
74. Yadav, S., Samadhiya, A., Kumar, A., et al. (2024). Nexus between Fintech, green finance and natural resources management: Transition of BRICS nation industries from resource curse to resource blessed sustainable economies. *Resources Policy, 91*, 104903. https://doi.org/10.1016/j.resourpol.2024.104903
75. Sadiq, M., Paramaiah, C., Joseph, R., et al. (2024). Role of fintech, green finance, and natural resource rents in sustainable climate change in China. Mediating role of environmental regulations and government interventions in the pre-post COVID eras. *Resources Policy, 88*, 104494. https://doi.org/10.1016/j.resourpol.2023.104494
76. Song, Y., & Hao, Y. (2024). Understanding the relationship between Fintech, natural resources, green finance, and environmental sustainability in China: A BARDL approach. *Resources Policy, 89*, 104608. https://doi.org/10.1016/j.resourpol.2023.104608
77. Tang, P., Ma, H., Sun, Y., & Xu, X. (2024). Exploring the role of Fintech, green finance and natural resources towards environmental sustainability: A study on ASEAN economies. *Resources Policy, 94*, 105115. https://doi.org/10.1016/j.resourpol.2024.105115
78. Wang, H., Jiao, S., Ge, C., & Sun, G. (2024). Corporate ESG rating divergence and excess stock returns. *Energy Economics, 129*, 107276. https://doi.org/10.1016/j.eneco.2023.107276

79. Liu, X., & Wan, D. (2023). Retail investor trading and ESG pricing in China. *Research in International Business and Finance, 65*, 101911. https://doi.org/10.1016/j.ribaf.2023.101911
80. Abudy, M. M., Gavious, I., & Shust, E. (2023). Does adopting voluntary ESG practices affect executive compensation? *Journal of International Financial Markets, Institutions and Money, 83*, 101718. https://doi.org/10.1016/j.intfin.2022.101718
81. Agnese, P., Battaglia, F., Busato, F., & Taddeo, S. (2023). ESG controversies and governance: Evidence from the banking industry. *Finance Research Letters, 53*, 103397. https://doi.org/10.1016/j.frl.2022.103397
82. Anita, M., Shveta, S., Yadav Surendra, S., & Arvind, M. (2023). When do ESG controversies reduce firm value in India? *Global Finance Journal, 55*, 100809. https://doi.org/10.1016/j.gfj.2023.100809
83. Bang, J., Ryu, D., & Webb, R. I. (2023). ESG controversy as a potential asset-pricing factor. *Finance Research Letters, 58*, 104315. https://doi.org/10.1016/j.frl.2023.104315
84. Xue, R., Wang, H., Yang, Y., et al. (2023). The adverse impact of corporate ESG controversies on sustainable investment. *Journal of Cleaner Production, 427*, 139237. https://doi.org/10.1016/j.jclepro.2023.139237
85. Wang, X., Wang, J., Guan, W., & Taghizadeh-Hesary, F. (2023). Role of ESG investments in achieving COP-26 targets. *Energy Economics, 123*, 106757. https://doi.org/10.1016/j.eneco.2023.106757
86. Cicchiello, A. F., Cotugno, M., & Foroni, C. (2023). Does competition affect ESG controversies? Evidence from the banking industry. *Finance Research Letters, 55*, 103972. https://doi.org/10.1016/j.frl.2023.103972
87. Bang, J., Ryu, D., & Yu, J. (2023). ESG controversies and investor trading behavior in the Korean market. *Finance Research Letters, 54*, 103750. https://doi.org/10.1016/j.frl.2023.103750
88. Xue, R., Wang, H., Yang, Y., et al. (2023). The adverse impact of corporate ESG controversies on sustainable investment. *Journal of Cleaner Production, 427*, 139237. https://doi.org/10.1016/j.jclepro.2023.139237
89. Gill, R. (2002). Change management—Or change leadership? *Journal of Change Management, 3*, 307–318. https://doi.org/10.1080/714023845
90. Pett, T., Haddad, G., & Nagpal, G. (2024). Does empowering leadership affect SMEs' sustainability performance through knowledge transfer? *European Management Journal.* https://doi.org/10.1016/J.EMJ.2024.03.008
91. Pantouvakis, A., & Vlachos, I. (2020). Talent and leadership effects on sustainable performance in the maritime industry. *Transportation Research Part D: Transport and Environment, 86*, 102440. https://doi.org/10.1016/j.trd.2020.102440
92. Cun, W. (2024). Business simulation analysis based on leadership development with crisis management using gaming techniques and machine learning model. *Entertainment Computing, 50*, 100698. https://doi.org/10.1016/J.ENTCOM.2024.100698
93. Alo, O., Ali, I., Zahoor, N., et al. (2023). Impression management and leadership in failing or failed business-to-business firms during and post-COVID-19: Empirical insights from Africa. *Industrial Marketing Management, 113*, 1–13. https://doi.org/10.1016/j.indmarman.2023.05.018
94. Qian, S., & Yu, W. (2024). Green finance and environmental, social, and governance performance. *International Review of Economics and Finance, 89*, 1185–1202. https://doi.org/10.1016/J.IREF.2023.08.017
95. Zhao, X., Zeng, B., Zhao, X., et al. (2024). Impact of green finance on green energy efficiency: A pathway to sustainable development in China. *Journal of Cleaner Production, 450*, 141943. https://doi.org/10.1016/j.jclepro.2024.141943
96. Yang, J., & Hui, N. (2024). How digital finance affects the sustainability of corporate green innovation. *Finance Research Letters, 63*, 105314. https://doi.org/10.1016/j.frl.2024.105314

97. Wei, H., Yue, G., & Khan, N. U. (2024). Uncovering the impact of Fintech, natural resources, green finance and green growth on Environment sustainability in BRICS: An MMQR analysis. *Resources Policy, 89*, 104515. https://doi.org/10.1016/j.resourpol.2023.104515
98. Ma, X., Liu, L., & Zhang, D. (2024). How green finance tools and electric vehicles minerals sustainability are related? *Resources Policy, 90*, 104799. https://doi.org/10.1016/j.resourpol.2024.104799
99. Chu, M., Li, B., Gu, W., & Dai, X. (2024). Role of green finance in enhancing the sustainability in the mining sector in Asia. *Resources Policy, 88*, 104473. https://doi.org/10.1016/j.resourpol.2023.104473
100. Xie, C., Ye, L., Zhong, N., & Wan, W. (2024). Impact of digital finance on corporate green innovation: Exploring role of land resource misallocation in China. *Resources Policy, 91*, 104920. https://doi.org/10.1016/J.RESOURPOL.2024.104920
101. Zhang, L., Sun, H., Pu, T., et al. (2024). Do green finance and hi-tech innovation facilitate sustainable development? Evidence from the Yangtze River Economic Belt. *Economic Analysis and Policy, 81*, 1430–1442. https://doi.org/10.1016/j.eap.2024.02.005
102. Zheng, J., Jiang, Y., Cui, Y., & Shen, Y. (2023). Green bond issuance and corporate ESG performance: Steps toward green and low-carbon development. *Research in International Business and Finance, 66*, 102007. https://doi.org/10.1016/j.ribaf.2023.102007
103. Souguir, Z., Lassoued, N., Khanchel, I., & Bouzgarrou, H. (2024). Environmental performance and corporate tax avoidance: Greenwashing policy or eco-responsibility? The moderating role of ownership structure. *Journal of Cleaner Production, 434*, 140152. https://doi.org/10.1016/J.JCLEPRO.2023.140152
104. Ijassi, W., Evrard, D., & Zwolinski, P. (2024). Development of a circularity design methodology for urban factories based on systemic thinking and stakeholders engagement. *Sustainable Production and Consumption, 46*, 600–616. https://doi.org/10.1016/j.spc.2024.02.031
105. Araújo Vila, N., & Otegui Carles, A. (2024). Sustainable tourism business. *Reference Module in Social Sciences*. https://doi.org/10.1016/B978-0-443-13701-3.00015-3
106. Welsh Government. (2023). *The business of becoming a sustainable nation*.
107. Huang, L. (2024). Green bonds and ESG investments: Catalysts for sustainable finance and green economic growth in resource-abundant economies. *Resources Policy, 91*, 104806. https://doi.org/10.1016/j.resourpol.2024.104806
108. Piwowar-Sulej, K., Malik, S., Shobande, O. A., et al. (2023). A contribution to sustainable human resource development in the era of the COVID-19 pandemic. *Journal of Business Ethics, 191*, 337. https://doi.org/10.1007/S10551-023-05456-3
109. Aguado-Correa, F., de la Vega-Jiménez, J. J., López-Jiménez, J. M., et al. (2023). Evaluation of non-financial information and its contribution to advancing the sustainable development goals within the Spanish banking sector. *European Research on Management and Business Economics, 29*, 100211. https://doi.org/10.1016/j.iedeen.2022.100211
110. Papandreou, A., Koundouri, P., & Papadaki, L. (2020). *Sustainable shipping: Levers of change*. Springer.
111. Wu, R., & Qin, Z. (2024). Asymmetric volatility spillovers among new energy, ESG, green bond and carbon markets. *Energy, 292*, 130504. https://doi.org/10.1016/j.energy.2024.130504
112. Scott, J. (2013). *The sustainable business: A practitioner's guide to achieving long-term profitability and competitiveness*. Routledge.
113. Isaak, R. (2002). The making of the ecopreneur. *Greener Management International, 38*, 81–91. https://doi.org/10.9774/GLEAF.3062.2002.SU.00009
114. Noei, S., Sargolzaei, A., Yen, K., et al. (2017). Ecopreneur selection using fuzzy similarity TOPSIS variants. *American Journal of Industrial and Business Management, 07*, 864–880. https://doi.org/10.4236/AJIBM.2017.77061
115. Schiemann, F., & Tietmeyer, R. (2022). ESG controversies, ESG disclosure and analyst forecast accuracy. *International Review of Financial Analysis, 84*, 102373. https://doi.org/10.1016/j.irfa.2022.102373

116. Ragazou, K., & Sklavos, G. (2021). *Circular economy as a footpath for regional development in European Union*. https://doi.org/10.33422/IVCSS.2020.05.152
117. Ragazou, K., Passas, I., & Garefalakis, A. (2022). It is time for anti-bribery: Financial institutions set the new strategic "roadmap" to mitigate illicit practices and corruption in the market. *Administrative Sciences, 12*, 166. https://doi.org/10.3390/ADMSCI12040166
118. Ragazou, K., Passas, I., Garefalakis, A., & Zopounidis, C. (2023). Business intelligence model empowering SMEs to make better decisions and enhance their competitive advantage. *Discover Analytics, 1*, 2. https://doi.org/10.1007/S44257-022-00002-3
119. Preetha, J., & Anand, V. A. (2024). Structural equation modelling (SEM) of employees' involvement towards strategic implementation of the total quality management (TQM) in Indian telecom industry limited, India during COVID pandemic. *Measurement: Sensors, 33*, 101134. https://doi.org/10.1016/j.measen.2024.101134
120. Schaltegger, S., & Sturm, A. (1989). *Okologieinduzierte ¨ Entscheidungsprobleme des managements. Ansatzpunkte zur Ausgestaltung von Instrumenten [Ecology induced management decision support. Starting points for instrument formation]*. University of Basel.
121. Schaltegger, S., & Wagner, M. (2005). Current trends in environmental cost accounting—And its interaction with ecoefficiency performance measurement and indicators. In *Implementing environmental management accounting: Status and challenges* (pp. 45–62). Springer.
122. Mudau, T. N., Cohen, J., & Papageorgiou, E. (2024). Determinants and consequences of routine and advanced use of business intelligence (BI) systems by management accountants. *Information and Management, 61*, 103888. https://doi.org/10.1016/j.im.2023.103888
123. Achmad, G. N., Yudaruddin, R., Nugroho, B. A., et al. (2023). Government support, eco-regulation and eco-innovation adoption in SMEs: The mediating role of eco-environmental. *Journal of Open Innovation: Technology, Market, and Complexity, 9*, 100158. https://doi.org/10.1016/J.JOITMC.2023.100158
124. Wang, H., Shen, H., & Li, S. (2023). ESG performance and stock price fragility. *Finance Research Letters, 56*, 104101. https://doi.org/10.1016/j.frl.2023.104101
125. Huang, L., & Chen, A. (2024). Family business succession and corporate ESG behavior. *Finance Research Letters, 60*, 104901. https://doi.org/10.1016/J.FRL.2023.104901
126. Giakoumelou, A., Salvi, A., Bekiros, S., & Onorato, G. (2024). ESG and FinTech funding in the EU. *Research in International Business and Finance, 69*, 102233. https://doi.org/10.1016/j.ribaf.2024.102233
127. Bockel, L., & Schiettecatte, L.-S. (2018). *Life cycle analysis and the carbon footprint of coffee value chains* (pp. 359–382). https://doi.org/10.19103/AS.2017.0022.20
128. Ragazou, K., Lemonakis, C., Passas, I., et al. (2024). ESG-driven ecopreneur selection in European financial institutions: Entropy and TOPSIS analysis. *Management Decision*. (in press). https://doi.org/10.1108/MD-12-2023-2425/FULL/XML
129. Ling, A., Li, J., Wen, L., & Zhang, Y. (2023). When trackers are aware of ESG: Do ESG ratings matter to tracking error portfolio performance? *Economic Modelling, 125*, 106346. https://doi.org/10.1016/j.econmod.2023.106346
130. Liu, Q., Chen, R., & Yu, Q. (2024). Improving sustainable performance of China's new energy industry through collaborative innovation network resilience. *Renewable and Sustainable Energy Reviews, 201*, 114625. https://doi.org/10.1016/J.RSER.2024.114625
131. Beloskar, V. D., Haldar, A., & Gupta, A. (2024). Gender equality and women's empowerment: A bibliometric review of the literature on SDG 5 through the management lens. *Journal of Business Research, 172*, 114442. https://doi.org/10.1016/J.JBUSRES.2023.114442
132. Basulo-Ribeiro, J., Pimentel, C., & Teixeira, L. (2024). What is known about smart ports around the world? A benchmarking study. *Procedia Computer Science, 232*, 1748–1758. https://doi.org/10.1016/J.PROCS.2024.01.173
133. Sun, Y., Zhao, D., & Cao, Y. (2024). The impact of ESG performance, reporting framework, and reporting assurance on the tone of ESG disclosures: Evidence from Chinese listed firms. *Journal of Cleaner Production*, 142698. https://doi.org/10.1016/J.JCLEPRO.2024.142698

134. Xiong, Z., Liu, J., Yan, F., & Shi, D. (2024). Corporate ESG performance when neighboring the Environmental Protection Agency. *Journal of Environmental Management, 349*, 119519. https://doi.org/10.1016/J.JENVMAN.2023.119519
135. Madison, N., & Schiehll, E. (2021). The effect of financial materiality on esg performance assessment. *Sustainability (Switzerland), 13*, 3652. https://doi.org/10.3390/SU13073652/S1
136. Naseer, M. M., Khan, M. A., Bagh, T., et al. (2024). Firm climate change risk and financial flexibility: Drivers of ESG performance and firm value. *Borsa Istanbul Review, 24*, 106–117. https://doi.org/10.1016/j.bir.2023.11.003
137. Wu, Z., Lin, S., Chen, T., et al. (2023). Does effective corporate governance mitigate the negative effect of ESG controversies on firm value? *Economic Analysis and Policy, 80*, 1772–1793. https://doi.org/10.1016/j.eap.2023.11.018
138. Wu, B., Gu, Q., Liu, Z., & Liu, J. (2023). Clustered institutional investors, shared ESG preferences and low-carbon innovation in family firm. *Technological Forecasting and Social Change, 194*, 122676. https://doi.org/10.1016/j.techfore.2023.122676
139. Chun, W., Wang, Z., & Gon Kim, H. (2024). Do environmental regulations drive MNEs' equity ownership? Considering the impact of exogenous shocks on MNEs' cross-border acquisitions. *Journal of Business Research, 174*, 114503. https://doi.org/10.1016/J.JBUSRES.2024.114503
140. de Almeida Vittori Ferreira, M., Morgado, C. d. R. V., & Estellita Lins, M. P. (2024). Organizations and stakeholders' roles and influence on implementing sustainability requirements in construction projects. *Heliyon, 10*, E23762. https://doi.org/10.1016/J.HELIYON.2023.E23762
141. Mozaffarian, D., Callahan, E. A., Glickman, D., & Maitin-Shepard, M. (2024). Developing a national nutrition policy strategy to advance cardiometabolic health and health equity. *Cell Metabolism, 36*, 651–654. https://doi.org/10.1016/J.CMET.2024.02.001
142. Haider, Z., Wang, Y., & Wang, Y. (2024). Family firm, financial constraint, and environmental preparedness: An international study. *Journal of International Financial Markets, Institutions and Money*, 101979. https://doi.org/10.1016/J.INTFIN.2024.101979
143. Sparacino, A., Merlino, V. M., Brun, F., et al. (2024). Corporate social responsibility communication from multinational chocolate companies. *Sustainable Futures, 7*, 100151. https://doi.org/10.1016/J.SFTR.2024.100151
144. Bennouri, M., Cozarenco, A., & Nyarko, S. A. (2024). Women on boards and performance trade-offs in social enterprises: Insights from microfinance. *Journal of Business Ethics, 190*, 165–198. https://doi.org/10.1007/S10551-023-05391-3
145. Cingranelli, D., & Filippov, M. (2018). Are human rights practices improving? *American Political Science Review, 112*, 1083–1089. https://doi.org/10.1017/S0003055418000254
146. Atenidegbe, O. F., & Mogaji, K. A. (2023). Modeling assessment of groundwater vulnerability to contamination risk in a typical basement terrain using TOPSIS-entropy developed vulnerability data mining technique. *Heliyon, 9*, e18371. https://doi.org/10.1016/j.heliyon.2023.e18371
147. Aras, G., Tezcan, N., Furtuna, O. K., & Kazak, E. H. (2017). Corporate sustainability measurement based on entropy weight and TOPSIS: A Turkish banking case study. *Meditari Accountancy Research, 25*, 391–413. https://doi.org/10.1108/MEDAR-11-2016-0100/FULL/HTML
148. Zhou, Y., Chang, F. J., Chang, L. C., & Herricks, E. (2024). Elevating urban sustainability: An intelligent framework for optimizing water-energy-food nexus synergies in metabolic landscapes. *Applied Energy, 360*, 122849. https://doi.org/10.1016/j.apenergy.2024.122849
149. Maia, R. G. T., & Junior, A. O. P. (2021). Eco-efficiency of the food and beverage industry from the perspective of sensitive indicators of the water-energy-food nexus. *Journal of Cleaner Production, 324*, 129283. https://doi.org/10.1016/j.jclepro.2021.129283
150. Kim, A. R., Lee, S. W., & Seo, Y. J. (2022). How to control and manage vessels' ballast water: The perspective of Korean shipping companies. *Marine Policy, 138*, 105007. https://doi.org/10.1016/j.marpol.2022.105007

151. Erhemjamts, O., Huang, K., & Tehranian, H. (2024). Climate risk, ESG performance, and ESG sentiment in US commercial banks. *Global Finance Journal, 59*, 100924. https://doi.org/10.1016/j.gfj.2023.100924
152. Delgado, A., Rodriguez, D. J., Amadei, C. A., & Makino, M. (2024). Water in circular economy and resilience (WICER) framework. *Utilities Policy, 87*, 101604. https://doi.org/10.1016/j.jup.2023.101604
153. Luhede, A., Yaqine, H., Bahmanbijari, R., et al. (2024). The value of information in water quality monitoring and management. *Ecological Economics, 219*, 108128. https://doi.org/10.1016/j.ecolecon.2024.108128
154. Liu, D., Gu, K., & Hu, W. (2023). ESG performance and stock idiosyncratic volatility. *Finance Research Letters, 58*, 104393. https://doi.org/10.1016/j.frl.2023.104393
155. Jang, G. Y., Kang, H. G., & Kim, W. (2022). Corporate executives' incentives and ESG performance. *Finance Research Letters, 49*, 103187. https://doi.org/10.1016/j.frl.2022.103187
156. Zheng, S., Zhang, Q., & Zhang, P. (2023). Can customer concentration affect corporate ESG performance? *Finance Research Letters, 58*, 104432. https://doi.org/10.1016/j.frl.2023.104432
157. He, F., Guo, X., & Yue, P. (2024). Media coverage and corporate ESG performance: Evidence from China. *International Review of Financial Analysis, 91*, 103003. https://doi.org/10.1016/J.IRFA.2023.103003
158. Alam, A., Banna, H., Alam, A. W., et al. (2024). Climate change and geopolitical conflicts: The role of ESG readiness. *Journal of Environmental Management, 353*, 120284. https://doi.org/10.1016/j.jenvman.2024.120284
159. Chen, Y., Ren, Y. S., Narayan, S., & Huynh, N. Q. A. (2024). Does climate risk impact firms' ESG performance? Evidence from China. *Economic Analysis and Policy, 81*, 683–695. https://doi.org/10.1016/j.eap.2023.12.028
160. Friede, G., Busch, T., & Bassen, A. (2015). ESG and financial performance: Aggregated evidence from more than 2000 empirical studies. *Journal of Sustainable Finance and Investment, 5*, 210–233. https://doi.org/10.1080/20430795.2015.1118917
161. Wang, S., & Esperança, J. P. (2023). Can digital transformation improve market and ESG performance? Evidence from Chinese SMEs. *Journal of Cleaner Production, 419*, 137980. https://doi.org/10.1016/j.jclepro.2023.137980
162. Tran, B. L., Chen, C. C., & Tseng, W. C. (2022). Causality between energy consumption and economic growth in the presence of GDP threshold effect: Evidence from OECD countries. *Energy, 251*, 123902. https://doi.org/10.1016/j.energy.2022.123902
163. Ruey-Jenn, H., Cho-Min, L., Chien-Ming, H., & Chun-Wei, L. (2024). The impact of firm risk on the value of cash holdings: The moderating role of corporate social responsibility. *Pacific Basin Finance Journal, 83*, 102270. https://doi.org/10.1016/J.PACFIN.2024.102270
164. Amankwaah, E., & Baidoo, N. O. (2023). Effect of firm size and corruption on financial challenges of savings and loans companies: Evidence from Ghana. *Social Sciences and Humanities Open, 8*, 100552. https://doi.org/10.1016/j.ssaho.2023.100552
165. Kong, D., Zhu, L., & Wang, X. (2022). Anti-corruption and CEO compensation: Evidence from a natural experiment in China. *Economic Modelling, 106*, 105697. https://doi.org/10.1016/j.econmod.2021.105697
166. Sarhan, A. A., & Gerged, A. M. (2023). Do corporate anti-bribery and corruption commitments enhance environmental management performance? The moderating role of corporate social responsibility accountability and executive compensation governance. *Journal of Environmental Management, 341*, 118063. https://doi.org/10.1016/j.jenvman.2023.118063
167. Cicchiello, A. F., Cotugno, M., & Foroni, C. (2023). Does competition affect ESG controversies? Evidence from the banking industry. *Finance Research Letters, 55*, 103972. https://doi.org/10.1016/J.FRL.2023.103972

168. DasGupta, R. (2022). Financial performance shortfall, ESG controversies, and ESG performance: Evidence from firms around the world. *Finance Research Letters, 46*, 102487. https://doi.org/10.1016/j.frl.2021.102487
169. Latapí, M., Davíðsdóttir, B., & Jóhannsdóttir, L. (2023). Drivers and barriers for the large-scale adoption of hydrogen fuel cells by Nordic shipping companies. *International Journal of Hydrogen Energy, 48*, 6099–6119. https://doi.org/10.1016/j.ijhydene.2022.11.108
170. Tchapchet Tchouto, J.-E. (2023). An empirical assessment on the leveraging evidence of economic complexity under environmental kuznets curve hypothesis: A comparative analysis between Nordic and Non-Nordic European countries. *Innovation and Green Development, 2*, 100074. https://doi.org/10.1016/j.igd.2023.100074
171. Lindman, Å., Ranängen, H., & Kauppila, O. (2020). Guiding corporate social responsibility practice for social license to operate: A Nordic mining perspective. *Extractive Industries and Society, 7*, 892–907. https://doi.org/10.1016/j.exis.2020.07.013
172. Albu, N., Albu, C. N., Apostol, O., & Cho, C. H. (2020). The past is never dead: The role of imprints in shaping social and environmental reporting in a post-communist context. *Accounting, Auditing and Accountability Journal, 34*, 1109–1136. https://doi.org/10.1108/AAAJ-08-2019-4131/FULL/HTML
173. Apostol, O. (2022). Accounting for anticorruption: Where are the social and environmental accounting scholars? *Social and Environmental Accountability Journal, 42*, 223–234. https://doi.org/10.1080/0969160X.2022.2146151
174. Xiang, J., Kong, D., & Yang, W. (2022). Minimum wages and non-listed firm's earnings management: Evidence from a geographic discontinuity design. *Journal of Accounting and Public Policy, 41*, 106938. https://doi.org/10.1016/j.jaccpubpol.2022.106938
175. Bergerès, A. S., d'Astous, P., & Dionne, G. (2015). Is there any dependence between consumer credit line utilization and default probability on a term loan? Evidence from bank-customer data. *Journal of Empirical Finance, 33*, 276–286. https://doi.org/10.1016/j.jempfin.2015.03.007
176. Bokhari, S. A. A., Ali, M., Albort-Morant, G., et al. (2023). Bridging the gap: The moderating roles of institutional quality and compliance on the link between CSR and financial performance. *Journal of Cleaner Production, 404*, 136897. https://doi.org/10.1016/j.jclepro.2023.136897
177. Guo, S., Wang, C. L., Hwang, S., et al. (2022). Doing bad by doing good? Corporate social responsibility fails when controversy arises. *Industrial Marketing Management, 106*, 1–13. https://doi.org/10.1016/j.indmarman.2022.07.009
178. Kashmiri, S., Gala, P., & Nicol, C. D. (2019). Seeking pleasure or avoiding pain: Influence of CEO regulatory focus on firms' advertising, R&D, and marketing controversies. *Journal of Business Research, 105*, 227–242. https://doi.org/10.1016/j.jbusres.2019.08.022
179. Tian, Y., Ang, J. S., Fu, P., et al. (2023). Does social trust mitigate insiders' opportunistic behaviors? Evidence from insider trading. *Global Finance Journal, 59*, 100907. https://doi.org/10.1016/j.gfj.2023.100907
180. Ryu, D., Yang, H., & Yu, J. (2022). Insider trading and information asymmetry: Evidence from the Korea exchange. *Emerging Markets Review, 51*, 100847. https://doi.org/10.1016/j.ememar.2021.100847
181. Hu, F., Kusnadi, Y., Wang, J., & Wang, Y. (2022). Insider trading restrictions and real activities earnings management: International evidence. *Journal of International Financial Markets, Institutions and Money, 80*, 101641. https://doi.org/10.1016/j.intfin.2022.101641
182. Song, F., & Thakor, A. (2022). Ethics, capital and talent competition in banking. *Journal of Financial Intermediation, 52*, 100963. https://doi.org/10.1016/j.jfi.2022.100963
183. Bag, S., Rahman, M. S., Srivastava, G., et al. (2023). Examining the role of virtue ethics and big data in enhancing viable, sustainable, and digital supply chain performance. *Technological Forecasting and Social Change, 186*, 122154. https://doi.org/10.1016/j.techfore.2022.122154

184. Shakil, M. H. (2021). Environmental, social and governance performance and financial risk: Moderating role of ESG controversies and board gender diversity. *Resources Policy, 72*, 102144. https://doi.org/10.1016/j.resourpol.2021.102144
185. Ran, C., & Zhang, Y. (2023). The driving force of carbon emissions reduction in China: Does green finance work. *Journal of Cleaner Production, 421*, 138502. https://doi.org/10.1016/j.jclepro.2023.138502
186. Wang, L., & Peng, K. (2023). Carbon reduction decision-making in supply chain under the pledge financing of carbon emission rights. *Journal of Cleaner Production, 428*, 139381. https://doi.org/10.1016/j.jclepro.2023.139381
187. Fraser, A., & Fiedler, T. (2023). Net-zero targets for investment portfolios: An analysis of financed emissions metrics. *Energy Economics, 126*, 106917. https://doi.org/10.1016/j.eneco.2023.106917
188. Kaur, H., Gupta, S., & Dhingra, A. (2023). Selection of solar panel using entropy TOPSIS technique. *Materials Today Proceedings, 59*, 100907. https://doi.org/10.1016/j.matpr.2023.02.034
189. Kaur Bagga, S., Gera, S., & Haque, S. N. (2023). The mediating role of organizational culture: Transformational leadership and change management in virtual teams. *Asia Pacific Management Review, 28*, 120–131. https://doi.org/10.1016/j.apmrv.2022.07.003
190. Zare Banadkouki, M. R. (2023). Selection of strategies to improve energy efficiency in industry: A hybrid approach using entropy weight method and fuzzy TOPSIS. *Energy, 279*, 128070. https://doi.org/10.1016/j.energy.2023.128070
191. Saraiva, P. (2023). On Shannon entropy and its applications. *Kuwait Journal of Science, 50*, 194–199. https://doi.org/10.1016/j.kjs.2023.05.004
192. Liu, Q., Ma, J., Zhao, X., et al. (2023). Voltage fault diagnosis and misdiagnosis analysis of battery systems using the modified Shannon entropy in real-world electric vehicles. *Journal of Energy Storage, 73*, 109287. https://doi.org/10.1016/j.est.2023.109287
193. Hwang, C.-L., & Yoon, K. (1981). *Multiple attribute decision making* (Vol. 186). Springer. https://doi.org/10.1007/978-3-642-48318-9
194. Treepongkaruna, S., Kyaw, K., & Jiraporn, P. (2022). Shareholder litigation rights and ESG controversies: A quasi-natural experiment. *International Review of Financial Analysis, 84*, 102396. https://doi.org/10.1016/j.irfa.2022.102396
195. Basnyat, B., Treue, T., Pokharel, R. K., et al. (2023). Conservation by corruption: The hidden yet regulated economy in Nepal's community forest timber sector. *Forest Policy and Economics, 149*, 102917. https://doi.org/10.1016/j.forpol.2023.102917
196. Harring, N. (2014). Corruption, inequalities and the perceived effectiveness of economic pro-environmental policy instruments: A European cross-national study. *Environmental Science & Policy, 39*, 119–128. https://doi.org/10.1016/j.envsci.2013.08.011
197. Rahman, M. H., Tanchangya, T., Rahman, J., et al. (2024). Corporate social responsibility and green financing behavior in Bangladesh: Towards sustainable tourism. *Innovation and Green Development, 3*, 100133. https://doi.org/10.1016/j.igd.2024.100133
198. Muttakin, M. B., Monem, R. M., Khan, A., & Subramaniam, N. (2015). Family firms, firm performance and political connections: Evidence from Bangladesh. *Journal of Contemporary Accounting and Economics, 11*, 215–230. https://doi.org/10.1016/j.jcae.2015.09.001
199. Adelina, C., & Archer, D. (2024). A situated urban political ecology of plastics recycling in Bangkok. *Cities, 145*, 104671. https://doi.org/10.1016/j.cities.2023.104671
200. Owsen, D. M. (1997). The FASB's executive stock option controversy supports increasing the research and teaching of political accounting. *Accounting, Organizations and Society, 22*, 611–615. https://doi.org/10.1016/s0361-3682(97)00012-3
201. Sun, T., & Wang, X. (2023). Adoption of financial inclusion in a world of depleting natural resources: The importance of information and communication technology in emerging economies. *Resources Policy, 85*, 103901. https://doi.org/10.1016/j.resourpol.2023.103901
202. Crawford, L., Ferguson, J., Helliar, C. V., & Power, D. M. (2014). Control over accounting standards within the European Union: The political controversy surrounding the adoption of

IFRS 8. *Critical Perspectives on Accounting, 25*, 304–318. https://doi.org/10.1016/j.cpa.2013.03.001
203. Li, X., Liu, X., Li, X., & Liu, X. (2005). Foreign direct investment and economic growth: An increasingly endogenous relationship. *World Development, 33*, 393–407.
204. Chen, G., & Wang, M. (2023). Stock market liberalization and earnings management: Evidence from the China–Hong Kong Stock Connects. *Finance Research Letters, 58*, 104417. https://doi.org/10.1016/j.frl.2023.104417
205. Kua, H. W., Teoh, W. S., Xu, X., et al. (2024). A review of glass recycling policies in Stockholm, Hong Kong SAR and Shanghai from a circular economy perspective. *Journal of Cleaner Production, 434*, 140068. https://doi.org/10.1016/j.jclepro.2023.140068
206. Onal, B., Petmezas, D., & Xiong, N. (2022). Societal equality sentiment and executive compensation. *Journal of Corporate Finance, 77*, 102314. https://doi.org/10.1016/j.jcorpfin.2022.102314
207. Hoang, T. G., & Phan, T. N. T. (2024). The emergence of Fintech entrepreneurship: Empowering the transition to the digital economy. *Reference Module in Social Sciences*. https://doi.org/10.1016/B978-0-443-13701-3.00163-8
208. Lucey, B. M., Kumar, S., & Sureka, R. (2023). Corruption in finance research: The state of art and future research agenda. *Journal of Economic Criminology, 1*, 100001. https://doi.org/10.1016/J.JECONC.2023.100001
209. Caporin, M., Natvik, G. J., Ravazzolo, F., & Santucci de Magistris, P. (2019). The bank-sovereign nexus: Evidence from a non-bailout episode. *Journal of Empirical Finance, 53*, 181–196. https://doi.org/10.1016/j.jempfin.2019.07.001
210. Canta, C., Nilsen, Ø. A., & Ulsaker, S. A. (2023). Competition and risk taking in local bank markets: Evidence from the business loans segment. *Journal of Empirical Finance, 73*, 153–169. https://doi.org/10.1016/j.jempfin.2023.06.003
211. Anagnostopoulos, Y., Husa, K. A., & Noikokyris, E. (2022). A three-phase comparative efficiency analysis of US and EU banks. *International Review of Economics and Finance, 81*, 113–127. https://doi.org/10.1016/j.iref.2022.04.009
212. Robinson, S. C. (2020). Trust, transparency, and openness: How inclusion of cultural values shapes Nordic national public policy strategies for artificial intelligence (AI). *Technology in Society, 63*, 101421. https://doi.org/10.1016/j.techsoc.2020.101421
213. Dobers, P., & Wolff, R. (2000). Competing with 'soft'issues–from managing the environment to sustainable business strategies. Wiley Online Library.
214. Gong, M., Yu, K., Xu, Z., et al. (2024). Unveiling complementarities between national sustainable development strategies through network analysis. *Journal of Environmental Management, 350*, 119531. https://doi.org/10.1016/j.jenvman.2023.119531
215. Song, W., Han, Y. H., & Sroufe, R. (2022). Substitution and complementarity dynamics in configurations of sustainable management practices. *International Journal of Operations and Production Management, 42*, 1711–1731. https://doi.org/10.1108/IJOPM-10-2021-0647
216. Vaupel, M., Bendig, D., Fischer-Kreer, D., & Brettel, M. (2023). The role of share repurchases for firms' social and environmental sustainability. *Journal of Business Ethics, 183*, 401–428. https://doi.org/10.1007/S10551-022-05076-3
217. Maranghi, S., Parisi, M. L., Facchini, A., et al. (2020). Integrating urban metabolism and life cycle assessment to analyse urban sustainability. *Ecological Indicators, 112*, 106074. https://doi.org/10.1016/j.ecolind.2020.106074
218. Elster, J., & Hylland, A. (1989). *Foundations of social choice theory*. CUP.
219. Xue, Y., & Lai, K. H. (2023). Responsible shipping for sustainable development: Adoption and performance value. *Transport Policy (Oxford), 130*, 89–99. https://doi.org/10.1016/j.tranpol.2022.11.007
220. Solow, R. M. (2009). Sustainability: An economist's perspective. In *Saving small Island developing states: Environmental and natural resource challenges* (pp. 86–93).
221. Tran, T. H. U., Lau, K. H., & Ong, C. E. (2021). Adoption of social sustainability practices in an emerging economy: Insights from Vietnamese handicraft organizations. *Sustainable Production and Consumption, 28*, 1204–1215. https://doi.org/10.1016/j.spc.2021.07.034

222. Jokinen, L., Mäkelä, M., Heikkilä, K., et al. (2022). Creating futures images for sustainable cruise ships: Insights on collaborative foresight for sustainability enhancement. *Futures, 135*, 102873. https://doi.org/10.1016/j.futures.2021.102873
223. Lozano, R., Lukman, R., Lozano, F. J., et al. (2013). Declarations for sustainability in higher education: Becoming better leaders, through addressing the university system. *Journal of Cleaner Production, 48*, 10–19. https://doi.org/10.1016/J.JCLEPRO.2011.10.006
224. Ahn, Y. (2022). A socio-cognitive model of sustainability performance: Linking CEO career experience, social ties, and attention breadth. *Journal of Business Ethics, 175*, 303–321. https://doi.org/10.1007/S10551-020-04651-W
225. Townsend, J., & Barrett, J. (2015). Exploring the applications of carbon footprinting towards sustainability at a UK university: Reporting and decision making. *Journal of Cleaner Production, 107*, 164–176. https://doi.org/10.1016/j.jclepro.2013.11.004
226. Gavana, G., Gottardo, P., & Moisello, A. M. (2017). The effect of equity and bond issues on sustainability disclosure. Family vs non-family Italian firms. *Social Responsibility Journal, 13*, 126–142. https://doi.org/10.1108/SRJ-05-2016-0066/FULL/HTML
227. Khatter, A., McGrath, M., Pyke, J., et al. (2019). Analysis of hotels' environmentally sustainable policies and practices: Sustainability and corporate social responsibility in hospitality and tourism. *International Journal of Contemporary Hospitality Management, 31*, 2394–2410. https://doi.org/10.1108/IJCHM-08-2018-0670/FULL/XML
228. Ding, C., Gao, X., & Xie, Z. (2024). Analysing the differential impact of the COVID-19 pandemic on the resilience of the tourism economy: A case study of the Chengdu-Chongqing urban agglomeration in China. *International Journal of Disaster Risk Reduction, 102*, 104255. https://doi.org/10.1016/j.ijdrr.2024.104255
229. Perramon, J., Bagur-Femenías, L., Alonso-Almeida, M. d. M., & Llach, J. (2024). Does the transition to a circular economy contribute to business resilience and transformation? Evidence from SMEs. *Journal of Cleaner Production, 453*, 142279. https://doi.org/10.1016/J.JCLEPRO.2024.142279
230. Zhang, Z., Liu, J., Xiao, C., & Chen, G. (2023). Making waves: Enhancing sustainability and resilience in coastal cities through the incorporation of seawater into urban metabolism. *Water Research, 242*, 120140. https://doi.org/10.1016/j.watres.2023.120140
231. Bontempi, A., Del Bene, D., & Di Felice, L. J. (2023). Counter-reporting sustainability from the bottom up: The case of the construction company WeBuild and dam-related conflicts. *Journal of Business Ethics, 182*, 7–32. https://doi.org/10.1007/S10551-021-04946-6
232. Savitz, A. W., & Weber, K. (2006). The triple bottom line: How today's best-run companies are achieving economic, social, and environmental success-and how you can too. *Journal Soundview Executive Book Summaries, 28*, 2–8.
233. Elkington, J. (1999). Cannibals with forks: The triple bottom line of 21st century business. *Alternatives Journal, 25*, 42.
234. Stubbs, W., & Cocklin, C. (2008). Conceptualizing a "sustainability business model". *Organization and Environment, 21*, 103–127. https://doi.org/10.1177/1086026608318042
235. Tang, M., Hong, J., Wang, X., & He, R. (2020). Sustainability accounting of neighborhood metabolism and its applications for urban renewal based on emergy analysis and SBM-DEA. *Journal of Environmental Management, 275*, 111177. https://doi.org/10.1016/j.jenvman.2020.111177
236. Passetti, E., & Rinaldi, L. (2020). Micro-processes of justification and critique in a water sustainability controversy: Examining the establishment of moral legitimacy through accounting. *British Accounting Review, 52*, 100907. https://doi.org/10.1016/j.bar.2020.100907
237. Padilla-Rivera, A., Hannouf, M., Assefa, G., & Gates, I. (2023). A systematic literature review on current application of life cycle sustainability assessment: A focus on economic dimension and emerging technologies. *Environmental Impact Assessment Review, 103*, 107268. https://doi.org/10.1016/j.eiar.2023.107268

SPRINGER NATURE

GPSR Compliance

The European Union's (EU) General Product Safety Regulation (GPSR) is a set of rules that requires consumer products to be safe and our obligations to ensure this.

If you have any concerns about our products, you can contact us on ProductSafety@springernature.com

In case Publisher is established outside the EU, the EU authorized representative is:

Springer Nature Customer Service Center GmbH
Europaplatz 3
69115 Heidelberg, Germany

The manufacturer's authorised representative in the EU is Springer Nature Customer Service Centre GmbH, Europaplatz 3, 69115 Heidelberg, Germany. If you have any concerns regarding our products, please contact ProductSafety@springernature.com

Printed and bound by CPI Group (UK) Ltd, Croydon, CR0 4YY
01/10/2026
02225982-0001